Circulating Load: Practical Mineral Processing Plant Design

Circulating Load: Practical Mineral Processing Plant Design

Editor

Arvind Kumar

Circulating Load: Practical Mineral Processing Plant Design

Edited by **Arvind Kumar**

Printed in 2017

ISBN: 978-1-68117-477-8

Library of Congress Control Number: 2015936594

© 2016 by
SCITUS Academics LLC,
616, Corporate Way, Suite 2, 4766,
Valley Cottage, NY 10989

www.scitusacademics.com

This book contains information obtained from highly regarded resources. Copyright for individual articles remains with the authors as indicated. All chapters are distributed under the terms of the Creative Commons Attribution License, which permits unrestricted use, distribution, and reproduction in any medium, provided the original author and source are credited.

Notice

Reasonable efforts have been made to publish reliable data and views articulated in the chapters are those of the individual contributors, and not necessarily those of the editors or publishers. Editors or publishers are not responsible for the accuracy of the information in the published chapters or consequences of their use. The publisher believes no responsibility for any damage or grievance to the persons or property arising out of the use of any materials, instructions, methods or thoughts in the book. The editors and the publisher have attempted to trace the copyright holders of all material reproduced in this publication and apologize to copyright holders if permission has not been obtained. If any copyright holder has not been acknowledged, please write to us so we may rectify.

Contents

	Preface vii	
Chapter 1	Factors That Influence Failure Behaviour and Remaining Useful Life of Mining Equipment Components 1	
	Mark Ho and Melinda Hodkiewicz	
Chapter 2	Effort Estimation for Design Activity in Power Plant Equipments 27	
	Parimal Kumar Giri	
Chapter 3	A Novel Property of DNA – As a Bioflotation Reagent in Mineral Processing 45	
	Balasubramanian Vasanthakumar, Honnavar Ravishankar, and Sankaran Subramanian	
Chapter 4	Acoustical Design of an Electrical Emergency Plant Using Sea Method 67	
	Evgeny Podzharov, José F. de la Mora Gálvez, and Jesus A. Alvarez Sanchez	
Chapter 5	Thermo-Dynamical Analysis on Electricity-Generation Subsystem of CAES Power Plant 79	
	Wenyi Liu, Gang Xu, and Yongping Yang	
Chapter 6	The Optimal Steam Pressure of Thermal Power Plant in a Given Load 91	
	Yong Hu, Ji-zhen Liu, De-liang Zeng, Wei Wang, and Ya-zhe Li	
Chapter 7	Practical Implementation of Safety Verification in LNG Production Facilities 105	
	Achint Rastogi and Hossam A. Gabbar	
Chapter 8	Characterisation and Pre-concentration of Chromite Values from Plant Tailings Using Floatex Density Separator 135	
	C. Raghu Kumar, Sunil Tripathy, and D.S. Rao	

Chapter 9	**Cottonseed Yield and its Quality as Affected by Mineral Fertilizers and Plant Growth Retardants** 151	

Zakaria M. Sawan

Chapter 10	**Process Adaption and Modifications of a Nutrient Removing Wastewater Treatment Plant in Sri Lanka Operated at Low Loading Conditions** 205	

Johanna Berg and Stig Morling

Chapter 11	**Mineral Industry in Egypt-Part I: Metallic Mineral Commodities ... 225**	

Abdel-Zaher M. Abouzeid and Abdel-Aziz M. Khalid

Chapter 12	**Study on Operating Characteristics of Power Plant with Dry and Wet Cooling Systems** 269	

Tao Tang, Jian-qun Xu, Sheng-xiang Jin, and Hong-qi Wei

Citations 289
Index 293

Preface

Minerals' processing plant design and operations from the mining industry's leading engineers, consultants, and operators. In addition to valuable guidance on overall project management, the papers address the design, optimization, and control of all related processes, including crushing and grinding, separation, flotation, pumping and material transport, pre-oxidation, extraction, and proper disposal of by-products and tailings. Mineral processing plants have distributed control systems and information management systems; this book also describes the current platforms and toolkits available for implementing these advanced data processing and control systems. Some applications to real mineral processing plants or laboratory/pilot scale set-ups highlight the benefits obtained with the techniques described in the book.

Editor

Chapter 1

Factors That Influence Failure Behaviour and Remaining Useful Life of Mining Equipment Components

Mark Ho and Melinda Hodkiewicz

The University of Western Australia, Australia

ABSTRACT

Mobile mining equipment often operates in harsh environments characterised by remote locations and highly variable rock and operating conditions. This research explores the hypothesis that the failure behaviour of mining equipment is influenced by the physical properties of the ore and waste. We describe a method of examining this relationship via data mining on maintenance records and apply it to the hydraulic cylinders of two classes of earthmoving mobile equipment. Failure data for the analysis are drawn from maintenance work orders from 14 sites mining for haematite iron, nickel sulphide, and coking and thermal coal. The results show that the distributions of the estimated life parameters for hydraulic cylinders on earthmoving equipment are distinctly different for haematite iron, coal, and nickel sulphide sites. Analysis of the relationship between selected physical properties identified the influence of rock impact hardness number, abrasion index, and absolute hardness of the ore as significant factors for these hydraulic cylinders. Their effects are significant when parameters are considered in combination, for example, rock impact hardness number and abrasion index, and vary according to the cylinder type and asset class. The engineering implications of these results are considered with respect to known failure modes of the cylinders.

INTRODUCTION

Reliability is of key importance in the mining industry, and there have been considerable efforts put in over the last 10 years to improve the reliability of assets. Inherent reliability is a function of design, but the achieved reliability is influenced by a variety of operating circumstances. These include organisational processes such as how the asset should be maintained, culture in how it is operated, site factors such as asset age profile and functional expectations, and environmental conditions primarily associated with ore type and location. The aim of this work is to test the hypothesis that the failure behaviour of components is influenced by the operating context, specifically the characteristics of the ore body, and to identify factors that might be driving that behaviour. This is of concern to the mining industry as many deposits of note are situated in locations with extreme environmental conditions such as dust and temperature.

BACKGROUND

Reliability databases are used in the oil and gas, nuclear, chemical, and electronics industries for reliability assessments at the asset design stage, for benchmarking, and for reliability improvement programs. No such database exists in the mining industry; however, since 2010, CRC Mining has sponsored a research project to develop a framework for the sharing of failure data across organisations. In many cases in the mining industry, operators have only one or two high production assets such as excavators, shovels, crushers, or mills. With these small populations and limited failure data, it is a challenge to identify failure modes and predict failure behaviour. This impacts the engineer's ability to develop an effective maintenance program and accurately predict lifecycle costs and availability.

A shared database provides access to a larger number of failure events, and engineers can examine a wider range of failure modes than those that they may have experienced at their site. Beyond the availability of compiled data and the availability of standard reliability statistics, the question arises as to how to use the data in site-specific reliability assessments. If data are sought for a single-asset class such as a make and model of haul truck in haematite iron ore, then is it appropriate to compile data from identical haul trucks moving gold, iron ore, nickel, coal, and other commodities? Or should there be a process to tailor the analysis depending on the particular commodity and other site-specific characteristics? If so, do we need to establish what effect(s) site-specific characteristics may have on the failure behaviour and determine how to enable the use of "compiled" failure data on specific sites? Can we then leverage the knowledge from a larger number of failures yet still be able to tailor the data using site-specific factors?

The proportional hazards model (PHM) is originally developed by Cox [1] to determine significant factors and their magnitude of impact on the measure of interest. In the field of reliability engineering, particularly for mining equipment, PHM modelling work has been performed to determine the effects of condition monitoring covariates on remaining useful life. Examples include studies where data have been used to identify significant condition monitoring covariates affecting the times to failure of transmissions [2] and wheel motors [3] on mine trucks, rail wagon bearings [4], circulating pumps in a petrochemical plant [5], and diesel engines [6]. The PHM has been also used to evaluate the effect of noncondition monitoring covariates on times to failure. Examples of these include evaluating the effect on times to failure of the following: cable material choice for power cables in underground loader hauler dumpers [7], a combination of internal (condition monitoring) and external (organisational and maintenance) covariates affecting hydraulic jack units in underground loader hauler dumpers [8], effect of design characteristics and the impact of strike action on aluminium reduction cells [9], and operating characteristics of hydrocarbon pipelines. Work using related methods (proportional covariate model) in conjunction with accelerated life tests has been used to estimate hazard rates in mechanical systems [10]. The use of the PHM will be adopted in this study.

Work with other modelling techniques has also been performed such as using linear or quadratic regression to determine the relationship between engine health and engine performance for gas turbines [11], the effect of design and operating conditions on times to failure for pumps in an oil refinery using MANOVA and multivariate statistics [12], and the effect of pit and operating characteristics on the incidence of tyre failure using the F-test [13].

From an engineering perspective, it is a plausible hypothesis that an asset operating in a highly silicified gold mine will demonstrate different failure behaviour than that of the same asset in a bituminous coal operation. However, there has been very little published academic work to look at this. In part, this can be explained by the challenges in obtaining data across a sufficiently large population of sites mining for different ore types. This study uses the data collected in the CRC Mining study to explore the influence of external covariates by analysing individual subsystems across similar assets operating in multiple and diverse environments.

The challenge of dealing with data from different sites is compounded by the generally poor quality of reporting on failures generally found in computerised maintenance management systems and the considerable time required to clean the data [14, 15]. While it remains a possibility that the large original equipment manufacturers, especially those engaged in maintenance and repair contracts, have the data to perform this analysis, any work that has been done is not in the public domain. Given the absence of published work, a particular focus of this paper is setting out an appropriate methodology for preparation and analysis of the data.

METHODOLOGY

This section sets out the approach adopted to test the hypothesis that the failure behaviour of assets is influenced by the operating context,

specifically the characteristics of the ore. The process involves the following key steps.
- Select asset class, subsystem, and components.
- Define "failure" and cleanse data.
- Identify and characterise covariates of potential interest.
- Examine characteristics of the dataset.
- Explore effects of covariates on failure behaviour.
- Select PHM models using model selection criteria.

Select Asset Class, Subsystem, and Components

With industry assistance, we compiled a confidential failure database. It includes data from 14 mine sites, 4 commodities, and 6 organisational entities. The largest set of data is for heavy mobile equipment, including front-end loaders (FELs), dozers, shovels, graders, scrapers, drills, and dump trucks. Some models of equipment are used across all of the different commodity groups and at many of the sites. Mobile equipment has a number of components with similar functions. An example is the hydraulic cylinders used on FELs and dozers in surface mining operations. The two classes of hydraulic cylinders of interest are the following: (1) lift cylinders used in lifting or lowering a boom to which the implement is attached and (2) tilt cylinders responsible for changing the angle of the implement.

Define Failure and Cleanse Data

For lift and tilt cylinders, events classified as failures include (1) major leaks preventing the equipment from further operation and (2) loss of function to causes other than leaks. In responding to these failure events, two assumptions are made, the first assumption is that repair restore the component to as good as new and that when a replacement cylinder is installed, a new (not refurbished) cylinder. In order to test whether the failures are independent and identically distributed, an examination of the cumulative failures versus time plots is performed. This is described in a subsequent section. Events involving these cylinders that are NOT classified as failures include (1) preventative replacement of cylinders (a suspension is recorded), (2) accidental damage (a suspension is recorded), (3) minor leaks under observation but not preventing further operation, and (4) repairs or adjustments to cylinder attachments such as mountings, pins, valves, hoses, and rods. In total, there were a total of 1342 records of interest, including 551 suspensions and 791 failure events.

Identify and Characterise the Covariates of Potential Interest

The external covariates examined in this paper include values that can be measured via analysis of data provided, or examining the conditions present at each mine site. The covariates can be broadly classified into two categories: characteristics of the mine and characteristics of the operation and maintenance of each piece of equipment. Covariates examined in this study originating from the characteristics of the mine include ore properties such as rock impact hardness number (RIHN), absolute hardness, density, abrasive index, and unconfined compressive strength, quartile indicators for concentrations of sodium, potassium and silicon compounds, and 3 ore-type indicators corresponding to each ore type. Covariates examined originating from the characteristics of the operation and maintenance include the equipment size, noncompliance to mechanical lubrication schedules, inspection intervals, and duty level of the equipment. It is important to note that this list of covariates is not a complete list of all possible covariates; however, the study was restricted to covariates whose values were directly observable via provided data or published reference material.

In instances where the exact value for that mine site was not known, a value in the middle of the range for that ore type was assumed for each group. Data were compiled (in order of preference) from the mine site's geotechnical database (where available), estimated parameters from equipment manufacturers, or commonly accepted ranges from published reference material [16].

Examine Characteristics of the Dataset

Examination of Weibull Parameters of Stratified Population

Table 1 shows the shape parameter (β) of the Weibull distribution for each population of FEL and dozer cylinders. It can be seen that the β values for lift cylinders for both types of equipment are similar for Fe and coal but distinctly different for Ni. For the tilt cylinders, the β values are similar for coal and nickel but slightly lower for Fe.

Table 1: Weibull parameters and statistics of dataset

Type	(Fe)	(coal)	(Ni)	Number of failures Fe/coal/Ni	Number of suspensions Fe/coal/Ni
FEL lift cylinders	1.36	1.4	1.95	33/50/8	41/30/14
FEL tilt cylinders	1.07	1.25	1.32	30/52/6	41/32/12
Dozer lift cylinders	1.67	1.77	1.22	42/349/26	23/242/25
Dozer tilt cylinders	1.1	1.36	1.3	40/146/12	11/58/22

Other items of note in Table 1 are the low population of failures and high ratio of suspensions to failures (2:1) for cylinders on both FELs and dozers in nickel sulphide operations. Cylinders for FELs in haematite operations also exhibit high suspension to failure ratios with 1.36 suspensions per failure. This results in wide confidence intervals when estimating parameters; however, this is somewhat mitigated by the fact that both the combined populations as well as most populations stratified by commodity all have statistically significant numbers of failures greater than 30. An exception to this is cylinders for FELs mining for nickel sulphide for which there are sparse failure data (8 lift cylinder failures and 6 tilt cylinder failures) which are outnumbered by suspensions (14 lift cylinder suspensions and 12 tilt cylinder suspensions).

Verification of Independent and Identically Distributed Data

Verification that datasets are independent and identically distributed (IID) is performed by plotting plots of times between consecutive failures and cumulative failure times against failure number [17]. Identical distributions can be verified by ensuring a single linear trend in the cumulative failure times.

Figures 1(a), 1(b), 1(c), and 1(d) show the cumulative failure times between consecutive failures. There are no significant changes in slope that would indicate an increasing or decreasing hazard rate. This absence of any significant change in slope of these curves does not guarantee that repairs are as good as new, but it is indicative of an absence of factors that indicate increasing or decreasing life of subsequent failures or lack of independence of the dataset.

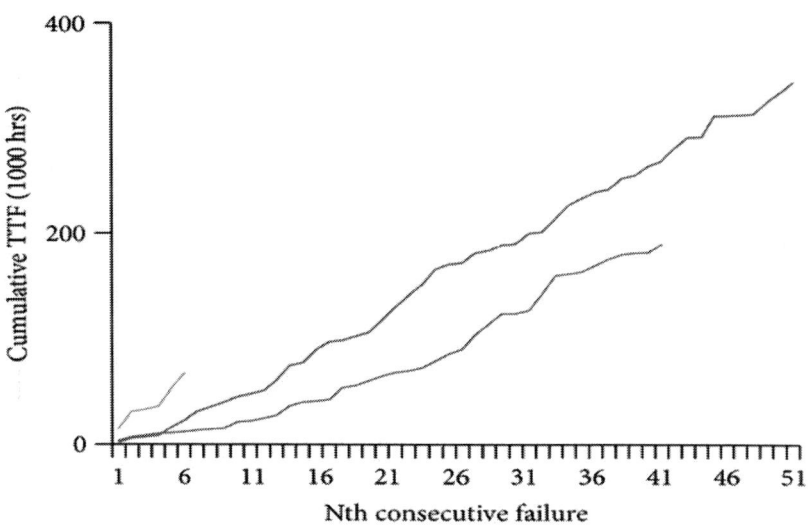

(a)

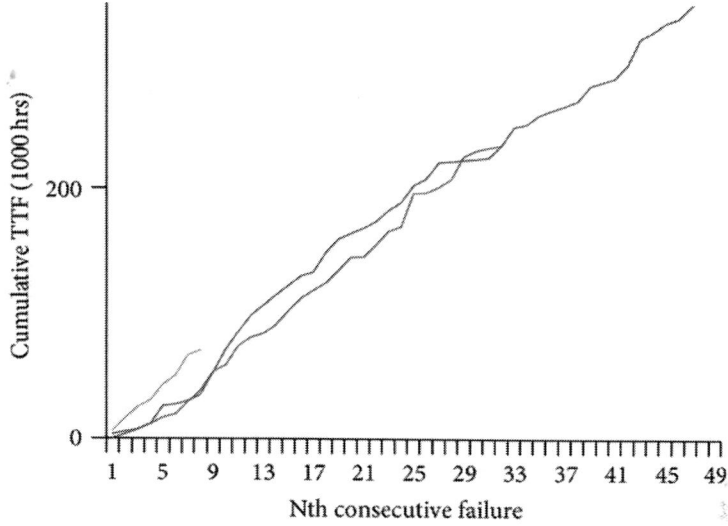

(b)

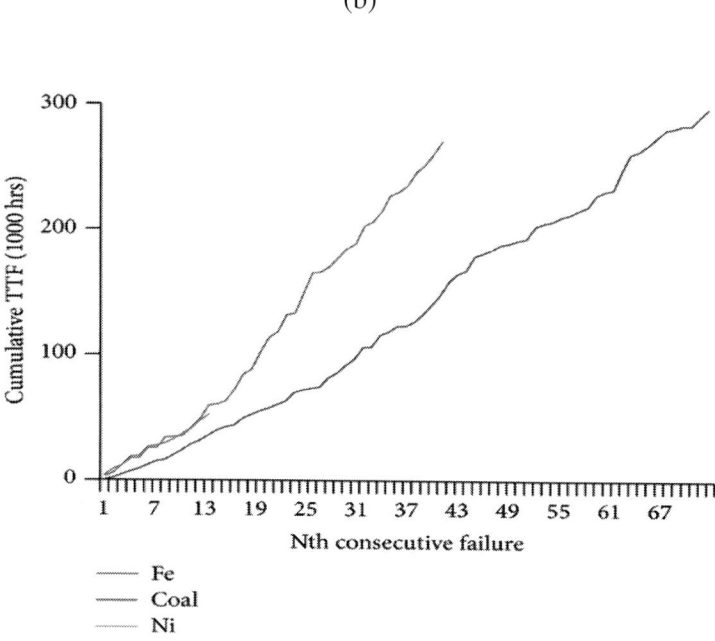

——— Fe
——— Coal
……… Ni

(c)

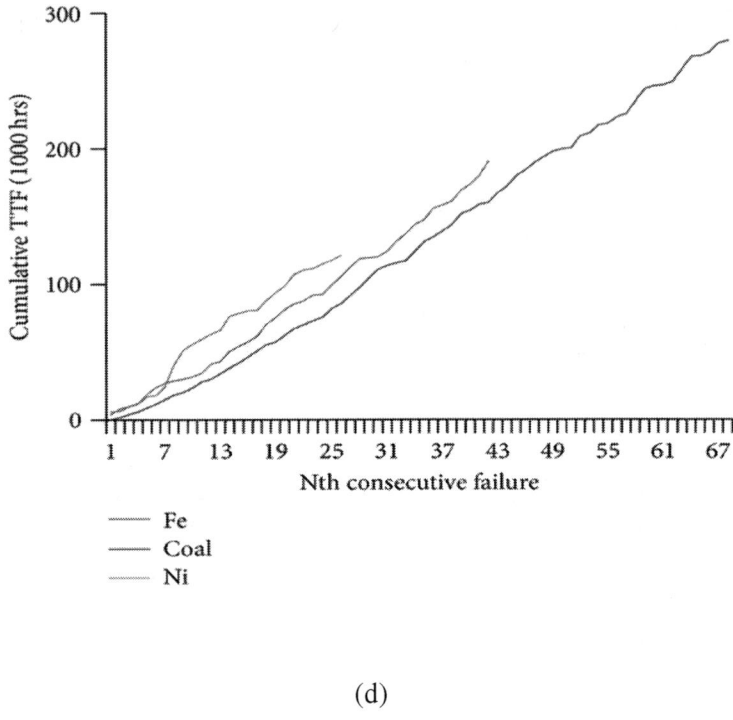

(d)

Figure 1: (a) FEL lift cylinder cumulative TTF. (b) FEL tilt cylinder cumulative TTF. (c) Dozer lift cylinder cumulative TTF. (d) Dozer tilt cylinder cumulative TTF.

Explore Effects of Covariates on Failure Behaviour

The relationships between ore characteristics and failure behaviour are explored statistically using proportional hazards modelling (PHM). For each cylinder type on each mobile equipment type, it is possible to construct a total of 15 one-covariate models (models with a single explanatory covariate) and 105 two-covariate models (models with two explanatory covariates). Due to the existence of only 3 ore types in this study, it is only possible to construct models with, at maximum, 2 covariates (corresponding to the 2 degrees of freedom available). This process was applied to both types of cylinders on each of the two types of mobile equipment leading to a total of 60 one-covariate models and 420 two-covariate models. The 480 models developed cover all possible combinations of the given covariates. All models were then tested for statistical accuracy and significance by the use of criteria in a model selection process.

The outputs from the proportional hazards model include the coefficients for each covariate, the standard error associated with each coefficient, and the baseline hazard rate. The coefficient of each covariate is proportional to the impact that covariate has on the hazard rate. Thus, positive coefficients indicate that the covariate increases the hazard rate while negative coefficients that indicate the covariate decreases the hazard rate. The baseline hazard rate can be modified by the calculated coefficients, and observed values of each covariate to obtain the modified hazard function and survival curves. A parametric fit is performed to determine the Weibull parameters for the theoretical distribution that best fits each modified survival curve. These are the predicted parameters that each PHM would predict for each commodity. The actual parameters are obtained by performing a standard Weibull analysis directly to the dataset of each commodity.

Select PHM Models Using Model Selection Criteria

In order to assess the suitability of each PHM model, we consider the following factors. The accuracy for the model is assessed by the average percentage deviation between the predicted and the actual MTTFs for all commodities. In this case, the lowest percentage deviation corresponds to the highest accuracy. Other measures of accuracy were considered, such as analysis of goodness of fit against the full Weibull curve for each commodity; however, the use of a point estimate (MTTF) comparison was selected due to its widespread use in industry. The significance of a covariate is assessed by comparing the magnitude of the coefficient with the magnitude of the standard error associated with that coefficient. Under this method, a higher ratio of coefficient to standard error indicates higher significance for that coefficient. The aim is to identify only models whose covariates most impact the hazard rate of the asset. Selection criteria for the models are as follows:
- high accuracy (less than 10% deviation between actual and predicted MTTF),

- each coefficient having a ratio at least 1.28 times the standard error (80% confidence),
- hypothes is test for proportionality failing to reject the proportional hazards assumption (PHA). (A value of 0.1 is used for this hypothesis test [18]. The null hypothesis states that the data are proportional, while the alternative hypothesis states that the data are not proportional.)

Due to the low complexity of the models (<2 covariates), it was not deemed necessary to consider the Akaike information criterion. Models that meet the model selection criteria are collated for similar subsystems across similar equipment types. Significant covariates or combinations of significant covariates are identified for further discussion and investigation.

RESULTS

Relationships between Ore Characteristics and Failure Behaviour

A visual representation of the cylinders life in operations mining for different commodities was obtained by the Bayesian inference [19]. These representations show that the distributions of the Weibull scale parameter (η) of different commodities are different. The FEL lift cylinders for nickel sulphide operations in Figure 2(a) are significantly different from those in coal and haematite operations. The FEL tilt cylinders for all commodities in Figure 2(b) are different from each other, and dozer tilt cylinders for haematite operations in Figure 3(b) are different from those in coal operations. It is important to note that these distributions are not frequency distributions of the times to failure.

Factors That Influence Failure Behaviour and Remaining Useful...

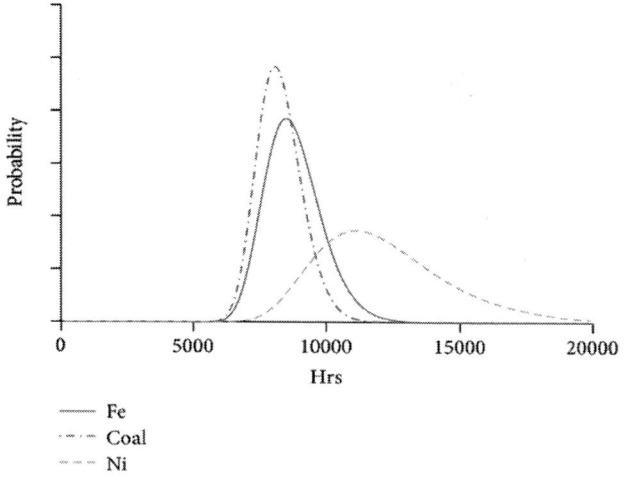

(a)

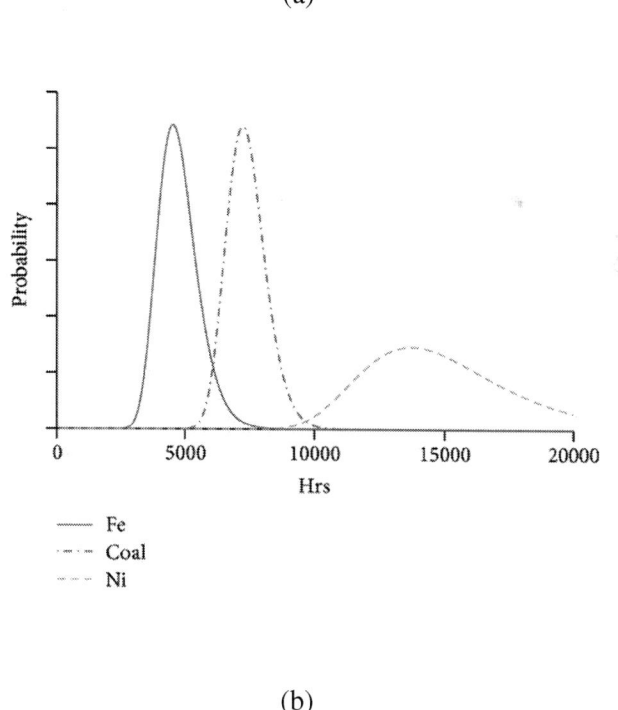

(b)

Figure 2: Distribution of η for FEL lift cylinders and FEL tilt cylinders.

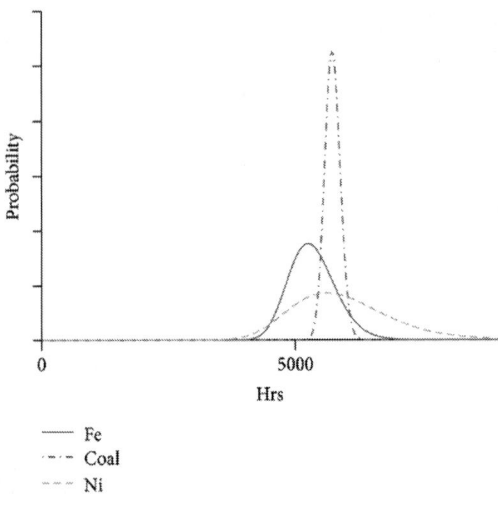

(a)

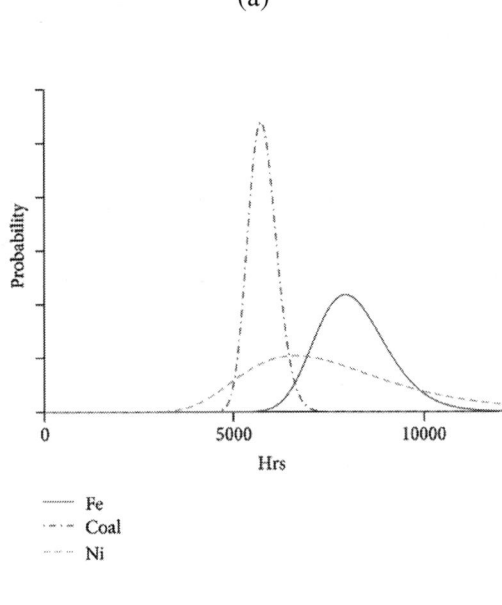

(b)

Figure 3: Distribution of η for dozer lift cylinders and dozer tilt cylinders.

Factors That Influence Failure Behaviour and Remaining Useful...

The PHM was applied to the cleansed data, and the selection process was used to identify appropriate models for lift and tilt cylinders on FELs and dozers. Table 2 shows a list of models that met the model selection criteria for accuracy and significance. A P value for the PHA hypothesis test is also included for each covariate. The P values were obtained using the cox.zph function in the survival package in R [20].

Table 2: Models meeting the model selection criteria for tilt and lift cylinders on dozers and FELs

Set	Covariates	Coefficient 1	Coefficient 2	Ratio Coef-SE1	Ratio Coef-SE2	PHA P value	MTTF deviation (%)
FEL-Lift	Abrasive index and absolute hardness	67.86	−0.22	1.77	1.8	0.597 0.587	3.81%
FEL-Lift	Nickel sulphide	−0.64	NA	1.60	NA	0.65	10%
FEL-Tilt	UCS and silicon %	−0.014	0.37	2.71	1.5	0.666 0.934	5.49%
FEL-Tilt	Abrasive index and absolute hardness	89.67	−0.29	2.16	2.21	0.11 0.11	5.3%
FEL-Tilt	Coal and nickel sulphide	0.42	−0.61	1.82	1.34	0.951 0.124	3.89%
Dozer-Lift	Absolute hardness and haematite	−0.046	2.82	2.84	3.02	<0.01 <0.01	6.3%
Dozer-Lift	Abrasive index and RIHN	8.25	−0.093	3.13	2.98	<0.01 <0.01	4.64%
Dozer-Lift	RIHN and haematite	−0.066	1.10	2.83	3.12	<0.01 <0.01	5.76%
Dozer-Tilt	No suitable models	NA	NA	NA	NA	NA	NA

Table 2 is organised as follows. Columns 3 and 4 show the coefficient values for the covariates listed in Column 2. Columns 5 and 6 show the ratios of coefficient magnitude to standard error with higher ratios indicating stronger significance. The P value for PHA hypothesis testing is shown in Column 7 with values greater than 0.1, indicating a failure to reject the PHA. The deviation between the actual and predicted value for the MTTF is shown in Column 8. The MTTF calculation takes into account both β and η values [21] as follows:

$$\text{MTTF} = \int_0^\infty R(t)\,dt = \eta\left(\frac{1}{\beta} + 1\right). \tag{1}$$

Table 2 shows that only 8 models from the 480 tested have significant relationships. These relationships apply to tilt and lift cylinders of FELs and lift cylinders of dozers. Most of the covariates are related to ore properties rather than ore types with absolute hardness and/or abrasion index appearing in four of the 8 models.

The three models for dozer lift cylinders (Table 2: Rows 7 to 9) violate the proportionality assumption with their P values for the PHA hypothesis test (Table 2: Column 7) less than 0.1.

Of the models for FEL tilt cylinders, the models for abrasive index and absolute hardness (Table 2—Row 5) and ore indicators for coal and nickel sulphide (Table 2: Row 6) have P values for the PHA hypothesis tests close to the threshold of 0.1 (0.11, 0.11, and 0.124, resp.). The covariate models created for dozer tilt cylinders did not yield any relationships of significance as no covariate selection combination was able to meet the accuracy criteria.

Table 3 summarises the list of parameters, and their occurrences in models are identified as significant. The "number of occurrences" is the number of times that covariate was present in models that met the model selection criteria. The "effect on hazard rate" uses the sign of the calculated coefficient to determine whether the covariate positively or adversely impacts the hazard rate.

Table 3: Covariates used in PHM that meet the model selection criteria

Parameter	Number of occurrences	Effect on hazard rate
Absolute hardness	2	Higher scratch resistance decreases the hazard rates

UCS	1	Higher UCS decreases the hazard rate
Si	1	Higher silicon % increases the hazard rate
Abrasive index	2	Higher AI increases the hazard rate
Ore indicators	3	Coal mines have higher hazard rates
		Nickel mines have lower hazard rates
		Haematite mines have lower hazard rates for FELs but higher hazard rates for dozers
Most common model		Absolute hardness and abrasive index

In order to determine the magnitude of impact of each covariate relative to every other covariate, we normalise the measured covariate values in order to obtain a unit standard deviation. Table 4 shows the normalised coefficients for all models meeting the model selection criteria. It can be seen that the normalised covariates with the largest impact per standard deviation change in covariate values are the abrasive index and absolute hardness with ranges of 5.86 to 7.66 and −6.05 to −7.9, respectively, This compares with covariates with less impact such as Si, with a normalised coefficient of 0.32.

Table 4: Normalised coefficients for models meeting model selection criteria

Set	Covariates	Normalised coefficient 1	Normalised coefficient 2
FEL-Lift	Abrasive index and absolute hardness	5.86	−6.05
FEL-Lift	Nickel sulphide	−0.64	NA
FEL-Tilt	UCS and silicon %	−0.59	0.32
FEL-Tilt	Abrasive index and absolute hardness	7.66	−7.9

| FEL-Tilt | Coal and nickel indicators | 0.42 | −0.61 |

DISCUSSION

Analysis of Distributions and Weibull Parameters Based on Commodity Type

Examination of the distributions shown in Figure 2(a) to Figure 3(b) shows the following.

- FEL Lift Cylinders. Cylinders in nickel sulphide operations have higher η and MTTF than those in haematite and coal operations. Coal and haematite operations have cylinders with comparable lifetime distributions.
- The high number of suspensions and low number of failures for FEL lift cylinders in nickel sulphide operations resulted in higher uncertainty in parameter estimation as evidenced by the wide dispersion of the distribution. Although haematite operations also have high numbers of suspensions, the 33 failures points were sufficient to give a distribution dispersion similar to that of the coal operations. This was also evident in FEL tilt cylinders.
- FEL Tilt Cylinders. Cylinders in nickel sulphide operations have higher η and MTTF than those in coal operations. Cylinders in haematite operations have the lowest η and MTTF out of all the three commodities.
- Dozer Lift Cylinders. Cylinders in haematite operations have lower η and MTTF than those in coal operations. The η and MTTF of nickel sulphide operations have a large variance and are not conclusively different from those in haematite or coal operations.
- Dozer Tilt Cylinders. Cylinders in haematite operations have higher η and MTTF than those in coal operations. The η and MTTF of nickel sulphide operations have a large variance and are not conclusively different from those in haematite or coal operations.

- The variance of the distributions for nickel sulphide and haematite iron operations is larger than that for coal operations. This is due to the availability of a larger dataset for coal operations, leading to higher certainty (probability) of parameter estimates as well as the high number of suspensions for nickel sulphide operations.

Tables 3 and 4 show that absolute hardness and abrasive index were parameters that had the highest rate of occurrence in all of the significant models. They were also the parameters with the highest impact per standard deviation change of observed covariate values.

Exploring the Relationship to Specific Physical Properties

The covariate models developed in this study identify, from a selected population of physical properties, those properties that impact the reliability of lift and tilt hydraulic cylinders on earthmoving mobile equipment. These properties are the abrasive index and absolute hardness of the mineral. To explore the relationship between what is observed in the data and what might be occurring in the physical environment, a review of common failure causes of hydraulic cylinders was conducted. The most common causes of cylinder failure (excluding rod breakage or alignment issues) are contamination, bearing and seal damage, chemical or heat degradation, and structural damage.

The coefficient calculated in this study for absolute hardness shows a decrease in hazard rate with absolute hardness. We postulate that the effect of a higher absolute hardness, which is the ability of a rock to resist scratching, decreases the ability of the material to create dust-sized particles. This results in a less dusty environment, which increases the reliability of a hydraulic cylinder due to less dust contamination of the hydraulic fluid or bearings. An increase in abrasiveness of the material appears to influence the reliability of the lift and tilt cylinders in some assets. In a high-dust environment, the ability of the particles to cause abrasive wear once in contact with the cylinder surfaces, bearings, or seals will accelerate deterioration.

The coefficient values and specific observed values for any commodity can be used to determine the magnitude of impact on the hazard rate for that commodity. In the case of the abrasive index and absolute hardness model, the coefficient values are 67.86 and −0.22, respectively (Table 2: Row 2, Columns 3 and 4). A FEL in a nickel sulphide mining operation with an abrasive index of 0.04 and absolute hardness of 15 would experience a modifying factor to the baseline hazard rate of 0.55 (45% lower than the baseline hazard rate). The baseline hazard rate represents the hazard rate of a fictional ore with an abrasive index of 0 and absolute hardness of 0. Similarly, a FEL in a coal mining operation with a low abrasion index of 0.01 and low absolute hardness of 2 would experience a modifying factor to the baseline hazard rate of 1.27 (27% higher than the baseline hazard rate). This implies that the lift cylinder of a FEL in a nickel mining operation experiences a hazard rate that is 43% of the hazard rate experienced by an equivalent FEL in a coal mining operation.

It is possible to apply the modifying factor for any specific commodity to obtain the commodity-specific hazard rate, survival curves, and associated Weibull parameters (β and η). The practical use of these coefficients is to extract operation-specific parameters from pooled data. These operation-specific parameters such as the MTTF, β, or η can be used to benchmark existing operations or provide insight when developing a mine site with differing operating conditions.

Organisational and Maintenance Properties

The analysis examined covariates representing the maintenance and operating conditions of the organisation including equipment size, equipment duty level, inspection intervals, and compliance to mechanical lubrication schedules. As a proxy, the scheduled compliance to a 250-hour engine service task was used. This maintenance work does not cover the cylinders specifically, but it is an indication of organisational commitment to scheduled maintenance. There are no direct data on scheduled maintenance compliance for the lift and tilt cylinders. Of these covariates only, the equipment duty level and compliance to mechanical lubrication schedules were found to be significant; however, they were not as significant as the ore properties. Equipment that was used for lower numbers of hours a day experienced higher hazard rates suggesting that deterioration due to time or exposure still occurs even when the equipment is not in use. Organisations with higher compliance to scheduled maintenance also experienced lower hazard rates. Further work in respect to developing and using organisational indicators as covariates is underway.

Sources of Uncertainty

There are a number of sources of uncertainty in this study due to the nature of failure data and also the simplifications inherent in compressing complex physical property distributions into a set of numbers. The original datasets contain data for all failure events, and there is considerable variability in the content of the fields and how data are represented. Extraction of failure and preventative replacement events was performed using an in-house data cleansing tool with the outputs cross-referenced against external data sources such as maintenance plans and external maintenance contracts. Incorporating data from these external data sources was done as follows.

Maintenance and repair contracts (MARCs) are specified time intervals wherein major maintenance and replacement activity is performed by an external contractor (most often the equipment vendor). Maintenance work performed under these contracts is recorded in the record system of the contractor rather than that of the mining company; it is therefore not included in the dataset. Data points immediately following a maintenance contract are removed as they also include times to failure/replacement of all maintenance work performed under the MARCs. Times between preventative replacements are compared against replacement intervals specified in the mining company's maintenance plan. Events marked as preventative replacements occurring significantly sooner than the planned replacement interval are treated as failures unless there is other evidence indicating that they were preventatively performed.

The dataset used includes 5 cases in which the number of failures is less than the number of suspensions. This can result in wider confidence intervals for both η and β values. Alternative methods of parameter estimation such as the Bayesian inference, as shown in Figure 2(a), or hypothesis testing can be used [22]. In addition, in two cases for nickel, there are low data populations resulting in poor fit. More data are being sought for nickel in order to improve confidence in estimated parameters.

Further Research

A work is currently underway to extend the range of components and asset classes under analysis. This will include mechanical, electrical/electronic, hydraulic, and structural subsystem components and asset classes from both fixed-plant and mobile equipments.

It was found that the assumption of proportionality was rejected for both types of cylinders for the dozer equipment class as shown in Table 2. Further work will be undertaken to more adequately model these assets by stratifying the asset class further or choosing an alternative modelling approach.

Approximations of dust levels at each mine site were inferred based on the ability of the material to resist fragmentation. This may not hold true due to other factors and dust-creation mechanisms such as the use of dust suppression, blasting techniques, presence of clay or other fine-grained materials, and nonimpact-related dust-creating mechanisms. Further work will be performed in assessing dust concentrations experienced at the rock interface on a mine site-specific basis. Additional environmental parameters such as average temperatures and chemical compositions of rocks will be added to determine whether other extreme environmental conditions have a significant influence on equipment failure.

It has been noted that the use of purely physical properties as covariates may not be sufficient to explain the failure behaviour of equipment. Additional covariates associated with organisational processes, site factors, and environmental conditions can be added as they become available. Organisational processes to be added include the culture in which the asset is operated (including functional role and level of loading) and site factors such as asset age profile and pit characteristics (e.g., single/double benching, blasting characteristics) that are controllable by the organisation. Maintenance factors to be added in future work include the effect of maintenance, inspection, or condition monitoring activities specific to the component. The framework supports the extension of the number of covariates to n-parameter models.

CONCLUSIONS

This study demonstrates that the distributions of the estimated life parameter for tilt and lift cylinders on dozers and FELs are distinctly different in location and shape for haematite iron, coal, and nickel sulphide operations. Further work is underway to extend the investigation to other components and assets.

The analysis uses proportional hazards modelling to extract context-specific information from compiled data. The choice of model and the covariates used in the model also give an insight into the factors that influence the reliability of the asset and their relative magnitudes. Analysis of the relationship between selected physical properties identified the influence of the abrasive index and absolute hardness of the ore as significant factors influencing the lifetime of tilt and lift cylinders.

It was also found that the assumption of proportionality of hazard rates was found not to hold for a specific asset class subsequent to performing the appropriate hypothesis test. This resulted in the rejecting of all models for cylinders on the dozer equipment class.

This work is significant because of the questions it raises about factors such as ore properties that may influence asset component failure. With this understanding comes the opportunity to reduce failure rates and tailor equipment selection, operation, and maintenance activities to specific sites. This will be explored in future papers.

REFERENCES

1. D. Cox, "Regression models and life-tables," Journal of the Royal Statistical Society, vol. 34, no. 2, pp. 187–220, 1972.
2. D. Lin, D. Banjevic, and A. K. S. Jardine, "Using principal components in a proportional hazards model with applications in condition-based maintenance," Journal of the Operational Research Society, vol. 57, no. 8, pp. 910–919, 2006.
3. A. K. S. Jardine, D. Banjevic, M. Wiseman, S. Buck, and T. Joseph, "Optimizing a mine haul truck wheel motors' condition monitoring program: use of proportional hazards modeling," Journal of Quality in Maintenance Engineering, vol. 7, no. 4, pp. 286–301, 2001.

4. A. Ghasemi and M. R. Hodkiewicz, "Estimating mean residual life for a case study of rail wagon bearings," IEEE Transactions on Reliability, vol. 61, no. 3, pp. 719–730, 2012.
5. P. J. Vlok, J. L. Coetzee, D. Banjevic, A. K. S. Jardine, and V. Makis, "Optimal component replacement decisions using vibration monitoring and the proportional-hazards model," Journal of the Operational Research Society, vol. 53, no. 2, pp. 193–202, 2002.
6. A. K. S. Jardine, P. Ralston, N. Reid, and J. Stafford, "Proportional hazards analysis of diesel engine failure data," Quality and Reliability Engineering International, vol. 5, no. 3, pp. 207–216, 1989.
7. D. Kumar, B. Klefsjö, and U. Kumar, "Reliability analysis of power transmission cables of electric mine loaders using the proportional hazards model," Reliability Engineering and System Safety, vol. 37, no. 3, pp. 217–222, 1992.
8. B. Ghodrati, U. Kumar, and F. Ahmadzadeh, "Remaining useful life estimation of mining equipment: a case study," in Proceedings of the International Symposium on Mine Planning and Equipment (MPES '12), 2012.
9. J. Kalbfleisch, "Measuring the impact of an intervention on equipment lives," Canadian Journal of Statistics, vol. 10, no. 4, pp. 237–259, 1982.
10. Y. Sun, L. Ma, J. Mathew, W. Wang, and S. Zhang, "Mechanical systems hazard estimation using condition monitoring," Mechanical Systems and Signal Processing, vol. 20, no. 5, pp. 1189–1201, 2006.
11. Y. G. Li and P. Nilkitsaranont, "Gas turbine performance prognostic for condition-based maintenance," Applied Energy, vol. 86, no. 10, pp. 2152–2161, 2009.
12. M. Braglia, G. Carmignani, M. Frosolini, and F. Zammori, "Data classification and MTBF prediction with a multivariate analysis approach," Reliability Engineering and System Safety, vol. 97, no. 1, pp. 27–35, 2012.
13. P. F. Knights and A. L. Boerner, "Statistical correlation of off-highway tire failures with openpit haulage routes," Mining Engineering, vol. 53, no. 8, pp. 51–56, 2001.

14. B. Veron and M. R. Hodkiewicz, "Rapid availability model development using time delay accounting," in Proceedings of the ICOMS Asset Management Conference, Sydney, Australia, 2009.
15. K. Unsworth, E. Adriasola, A. Johnston-Billings, A. Dmitrieva, and M. Hodkiewicz, "Goal hierarchy: improving asset data quality by improving motivation," Reliability Engineering and System Safety, vol. 96, no. 11, pp. 1474–1481, 2011.
16. G. Australia, "Geochemical Atlas of Australia," 2013, http://www.ga.gov.au/energy/projects/national-geochemical-survey/atlas.html.
17. D. M. Louit, R. Pascual, and A. K. S. Jardine, "A practical procedure for the selection of time-to-failure models based on the assessment of trends in maintenance data," Reliability Engineering and System Safety, vol. 94, no. 10, pp. 1618–1628, 2009.
18. P. M. Grambsch and T. M. Therneau, "Proportional hazards tests and diagnostics based on weighted residuals," Biometrika, vol. 81, no. 3, pp. 515–526, 1994.
19. M. Ho and K. Shen, "Analysis of failure data using proportional hazards and Bayesian inference," inProceedings of the 6th International Meeting on Mine Equipment Maintenance Meeting (MANTEMIN '11), Antofagasta, Chile, 2011.
20. "The Comprehensive R Archive Network," 2013, http://cran.r-project.org/.
21. M. Rausand and A. Hoyland, System Reliability Theory: Models, Statistical Methods and Applications, John Wiley & Sons, Hoboken, NJ, USA, 2nd edition, 2004.
22. J. Ansell and M. Phillips, Practical Methods for Reliability Analysis, Oxford Statistical Science Series, O.S. Publications, 1994.

Chapter 2

Effort Estimation for Design Activity in Power Plant Equipments

Parimal Kumar Giri

APEX Institute of Technology & Management Pahala, Bhubaneswar, India

ABSTRACT

This Software for Design Activity in power plants Equipments helps the power plant engineers and managers to manage the development and design activities of equipments in the field of power plants. This paper is basically concerned with the computerization of the design activity of Condenser, vital equipment in Heat Exchanger Unit (HEU) of Thermal power plant required for condensing the steam and for further reclaimable purposes to achieve economy. This software will also provide facilities to maintain user profile and the respective work details. The study use a developed model which estimates software effort by studying and analyzing small and medium scale application software.

INTRODUCTION

Design of power plant equipments, the sizing of equipments are reckoned for the economic and optimum design. So, software has been developed to this effect. This paper will help in finalizing optimum equipments size for tendering and in turn winning orders by the company from the customers. It will provide facilities to maintain equipment profile including all the parameters and the work details required to compute "PRESSURE DROP" parameter for the power equipment like "CONDENSER". A database is maintained with an aim to keep the records of all those employees who have attempted to access the software. For security purposes a password is also maintained for the software [1].

TYPES OF SOFTWARE

The number of different types of software components for the development of life cycle and associate with the different means of verification and validation with them are mentioned bellow. There are two main different types are new and existing software.

New software: all software written specifically for the application; Existing accessible software: typically software from a similar application that is to be reused and for which all the documentation is available; Existing proprietary software: typically a commercial product or software from another application that meets all or some of the current application requirements but for which little documentation is available; Configurable software: typically software that already exists but is configured for the specific application using data or an application specific input language.

Equipment Management

The equipments related information, of different phases, would be stored in separate databases [2]. The usage of Notepad files has also been made to store the frequently accessible data. Design engineering or design authority organization, represented by power plant technical services or power plant design engineering. Responsible for:
- Data collection, review, validation and input.
- Component design basis verification.
- Data entry organization.
- Data update procedures.
- Data update request.
- Software Quality Assurance (SQA).

Software Quality Assurance (SQA) also addresses software and hardware testing, and the concept of verification, validation and certification (V, V & C). Software testing and alignment with the appropriate software version and testing phase is tracked as well as interface relationships.

Software Lifecycle

The Software lifecycle is determined by a combination of technology lifespan and cost factors. Obsolescence for software occurs in different ways, and can be handled by upgrade (enhancement) or eventual replacement with more advanced software. System maintenance by vendor or inhouse technical staff can have a great influence on the useful life of a given solution.

Software Testing

After development, software is tested in several regimes: Unit testing, system testing, and user testing. If changes were made to existing software in a production environment, regression testing to ensure the changes do not impact other applications will be conducted also.

GENERAL PERCEPTION OF THE POWER PLANT

In general power plant consists of some major equipment such as boiler, turbine, condenser and generators. This paper deals with the development process of specific equipment CONDENSER with the help of computer applications. Initially the steam from the boiler equipment is allowed to fall on the blades of the turbine. This turbine consists of a rotor which is then further connected to the rotor of the generator. As soon as steam rotates the blades of the turbine, it further gives motion to the rotor which then further rotates the rotor of the generator. The generator, in addition to the rotor, consists of a rectangular coil which then rotates due to the circular motion of the rotor. The coil rotates in the magnetic field and due to this rotation induces current in it. The current thus generated above is then supplied to the electricity grid which is then distributed to different locations of the city. The basic anatomy of power plants has been elucidated in the Figure 1 given above.

OBJECTIVE OF THE SOFTWARE

The Design Activity Software helps design manager to manage resources anywhere (inside or outside the organization) using Internet and gives the liability to enhance the computing using different technologies [2].

- To meet the needs of the Heat Exchanger Unit (HXE).
- To develop a user-friendly interface for the engineers at power plants that could provide them an ease to understand and to execute the entire design process of huge plant equipments like Condensers.
- To find ways for human resources to "add value" to a business.

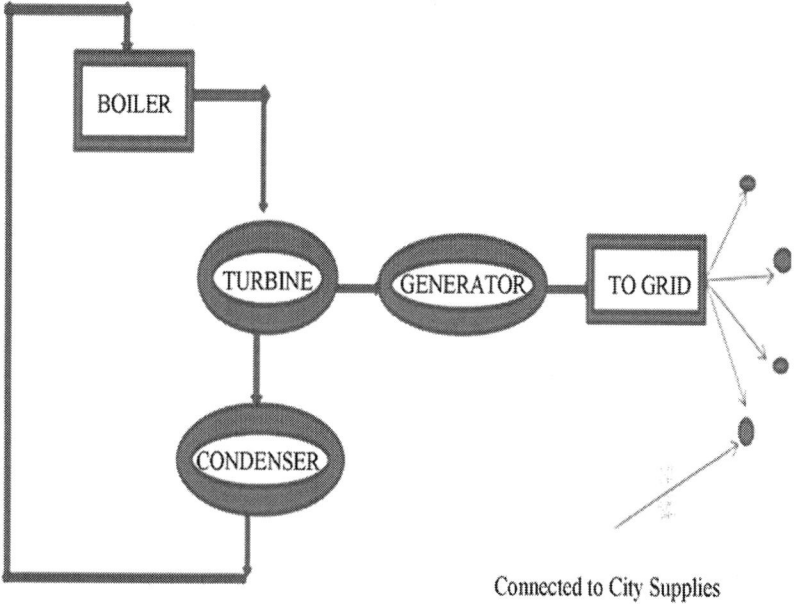

Figure 1: Basic anatomy of a power plant.

There are numerous pieces of equipment in a power plant, but the maintenance requirements are different for different equipment. Making condition based maintenance in a power plant is a systemic and integrative maintenance management activity which needs a special maintenance decision support system. As a main developing direction of decision support system, intelligent decision support system has got a comprehensive and successful application in many domains.

NEED FOR NEW SYSTEM

The newly proposed system is based on a user friendly model i.e. the entire interface, the controls and the handling, everything is very easy to learn and master. This model takes it closer to a utility. That means more reach, better performance, predictability, lower cost of ownership and all in all, great peace of mind from all business related problems. The new system has been so designed that it totally works on the user requirement. The objectives that are decided for the proposed system are:

- To computerize the existing system.
- Providing easy and interactive data entry interface.
- Providing effective calculation mechanism.
- Providing a way for analytic report generation.

THREE MAJOR MODULES INVOLVED

Sizing of Equipments

A program has been prepared to calculate the size of equipment considering the customer requirements. As customer requirements may vary from customer to customer so the first module plays an imperative role in determining the apt parameters required for the design phase of the equipment. Considering all the requirements, the best design is made and then the major parameter "pressure drop" will come into existence.

Since any condenser design is a balance of tube surface, tube length and cooling water quantity, it is the space allowed for the condenser that actually is the deciding factor in the design [1].

Generation of Performance Curve

Using the concepts of programming we'll generate the performance curve. This curve basically reflects the accuracy of the equipment. Once we have designed a condenser at the design point (i.e. the design condition) then our next task is to analyze the behavior of condenser under different conditions.

The Heat Exchange Institute (HEI) Standards for steam surface are used to design and predict the performance of the condensers for power plant applications [3, 4]. The paper suggests methodology to be followed for examining the performance using the HEI standard values.

Pressure Drop Computation

In this, basically the difference of pressure between two terminal points of any equipment is reckoned. This is the phase which is actually carried out at different sites that too under different temperature conditions.

COMPARISON BETWEEN THE EXISTING SYSTEM AND THE NEW SYSTEM

Existing System

The existing system is completely manual system which is little bit electronic gadget oriented as it involves the utilization of calculators. This system is very slow and doesn't handle all the things that are needed to be maintained in the organization. So, the calculations were to be done manually. Basically, the existing one is more error prone as the user is actually open to the commitment of certain human mistakes. The errors could be inefficiency on jotting down the important parameters, wrong calculation or mismanagement in plotting the graph related to the performance of the equipments.

All these shortcomings may lead to certain drastic changes and variations in the actual values of area of different equipments, may affect the analysis phase of equipments where their performance, under different conditions, is judged. As a result all these failures act as an impediment to the "success rate" of the company which is of the prime importance.

Earlier people used to rely on "punching card system" but as punch card workers learned more they were trusted with more complex tasks, and gradually moved toward the design of new procedures and wiring schemes to produce additional reports or tackle new jobs [3]. Figure 2 is Obsolete Punch Card System used in thermal power plants [5].

New System

On the contrary, the proposed system is made using VB.net as front end which is powerful Graphical User Interface tool. So anyone who doesn't have a thorough knowledge of computing can use the system. The proposed system manages all aspects of the development process of condensers that is all the information required in the different phase's right from sizing the equipment till the phase carried out at site after the designing of condenser [4]. It maintains the database of users (who intend to access he system), correction factors and input values.

SYSTEM ANALYSIS PHASE

System analysis refers to the process of examining a solution with the intent of improving it through better procedures and methods. System design is the process of planning a new system to either replace or complement an existing system. But before any planning is done, the old system must be thoroughly understood and the requirements determined. System analysis is therefore, the process of gathering and interpreting facts, and using the information to re-connect improvements in the system.

System analysis is conducted with the following objectives in mind:
- Identifying the customer needs.
- Evaluate the system concept for feasibility.
- Perform economic and technical analysis.
- Allocate functions to hardware, software, people, database and other system elements.
- Establish cost and schedule constraints.
- Create a system definition that forms the foundation for all the subsequent engineering work.

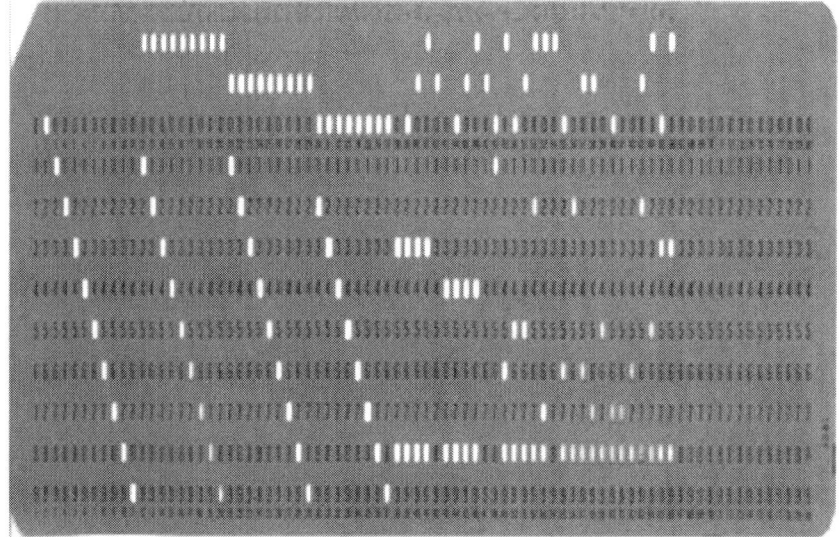

Figure 2: Obsolete punch card system used in thermal power plants.

Requirement Analysis

At the heart of the system analysis is a detailed understanding of all-important facts of the communication aspects between our company's internal networks. For this reason, the process of acquiring this information is often termed as detailed investigation. Analyst working closely with our clients, employees and managers has studied the business process. On the basis of the above requirements the capabilities to be achieved from the system are specified which lays a solid foundation for the system design of our module.

System Design

In the past few years, power plants are constructed with higher parameter, larger size and more automated with power industry developing quickly in China, which makes the system of a power plant become more complex and contain more equipment in quantity and type [6,7]. In order to ensure security and economy of production, power plants need to improve maintenance function in several ways:

- Finding out failure status of maintenance object and maintenance effect.
- Selecting right maintenance policy to achieve expected maintenance effect.
- Establishing quantitative models that quantify index and result in the maintenance policy.
- Monitoring operation status and health status of equipment in the system, to effectively control occurrence and development of failure.
- Making maintenance decision based on current information to effectively prevent occurrence and development of failure, ensure the security of equipment and personnel, and reduce economic lost caused by failure.
- Making maintenance decision and optimization with computer aid for operation and maintenance staff.

EQUIPMENT MECHANISM

Steam available at the turbine, with respect to Figure 1, consists of heat energy [5]. By the time the steam comes out of the turbine at the last stage (after travelling through the turbine) and is ready to enter the Condenser, the steam is at very reduced pressure called as "sub-atmospheric stage" i.e. a stage with pressure less than sub-atmospheric pressure. Since steam is at sub-atmospheric pressure so we need a vessel which is at the sub-atmospheric pressure.

Now, the exhaust of turbine is connected to the equipment or vessel called as "CONDENSER". It is that power plant equipment in which vacuum is created whose pressure is less than that of the exhaust or residue steam coming from the turbine [8]. This pressure within the vessel is created by other associated equipments called as "Ejectors". Finally, the condensed steam is fed to the boiler. After this only the latent heat is added to the converted steam.

Refer to Figure 3. Condensate flows in the shell side of the condenser and steam is condensed by the cooling water. Vacuum in the surface condenser i.e. turbine exhaust vacuum is controlled and maintained by vacuum Ejector system of the surface condenser [4]. Figure 4 shows the layout of a typical surface condenser used in steam turbines.

SOFTWARE DESIGN MODULES

This paper emphasis on a system that has the following modules designed from the analysis and design of the system:

Module 1—Sizing of Equipment

The condenser is the key component for the evaporative cooling system of hydro-generators. And the over-all heat transfer coefficient is primary standard to judge the condenser's performance [9]. In order to more comprehensively understand the evaporative cooling system and evaluate the condenser's work efficiency, it is important to calculate the over-all heat transfer coefficient of the condenser. In our works the over-all heat transfer was calculated in each stage and a kind of calculation methods was brought forward, thus the over-all heat transfer which is a value related to a process was proved.

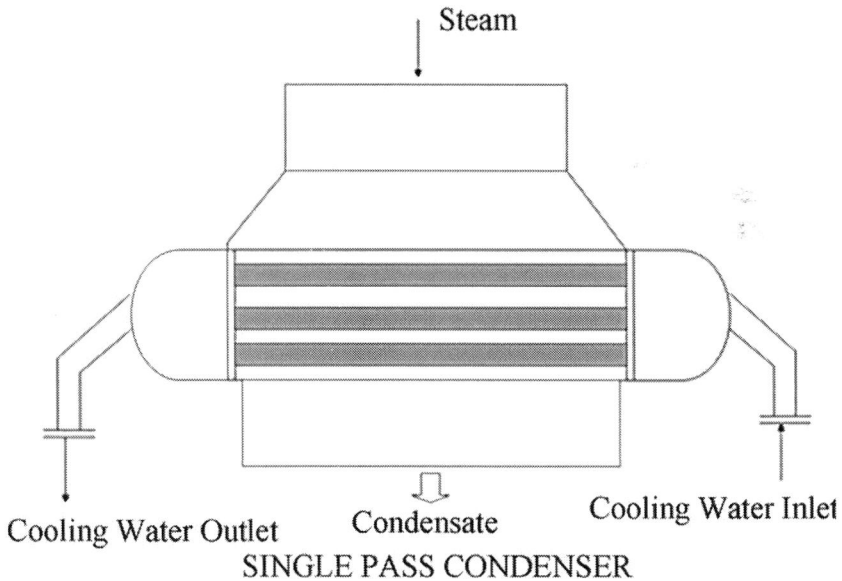

Figure 3: Schematic diagram of single phase condenser.

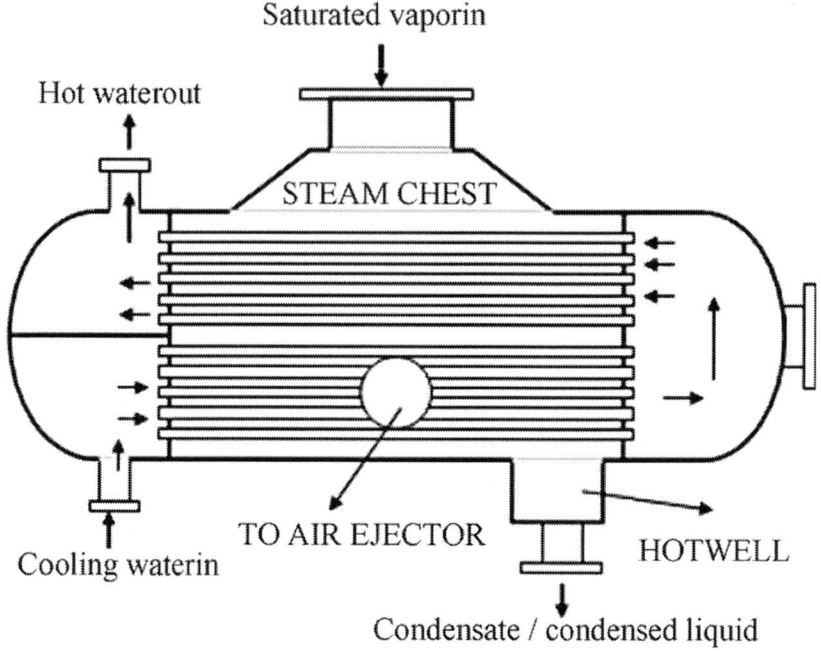

Figure 4: Surface condenser layout.

User is provided such an interface in which he is provided with three sub modules [10]:

- Design Information;
- Thermal Calculation; and
- Area of Condenser.

In this way, the application asks the programmer to enter the input information i.e. the parameters as provided by the customer of the Condenser. This information is termed as Design Information. Next, the calculations are carried out on the basis of the formulae as per HEI standards which are already fed in the programs in back end [3]. This phase comprises of Thermal Calculation part. Finally, once the programme is ready with all the input values, he can now find the Area of Condenser as per the given set of values by the customer.

Module 2—Performance Curves

Once the programmer has analyzed that what should be the area of Condenser for the given set of values now, he can accordingly analyze what exactly will be the performance of this condenser designed under different circumstances. The programmer can determine how condenser would perform at different values of Heat Load ("Q"—the amount of heat a Condenser can condense) and Back Pressure ("Ps"— Condenser Pressure which is less than atmospheric pressure). These two parameters are the most imperative parameters included in the Performance Analysis phase of the Condenser. User is provided with an interface in which he is required to enter the information as per the customer's SRS and then he is provided with the comparative study of the Condenser at different "Q" and "Ps" with the help of graphical representation as shown in the Figures 5 and 6.

To enhance the performance of the condenser we can also use 3D Modeling capabilities [8] in addition to the technique specified in this section.

Figure 5 shows the performance of condenser at three different temperatures 15 degC, 20 degC and 25 degC.

Figure 6 shows the performance of condenser at four different temperatures 15 degC, 20 degC, 25 degC, 30 degc, 35 degC and 40 degC.

Module 3—Pressure Drop Computation

This is the last phase of the designing process of the condenser. By this point of time we are ready with our Condenser and now it's the time for the testing of Condenser at the site where it is to be installed. At this location, we reckon an important factor named "Pressure Drop" which is basically the difference between the pressure at the inlet end and the outlet end of the Condenser [11].

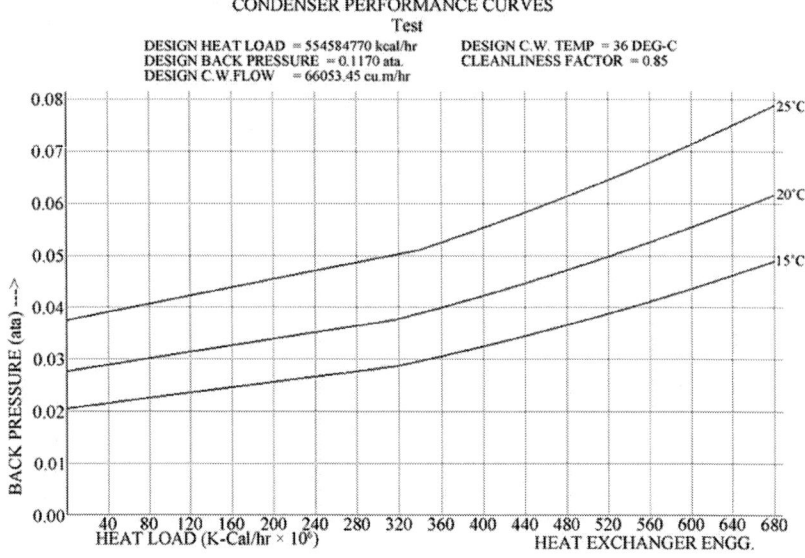

Figure 5: Condenser performance curves.

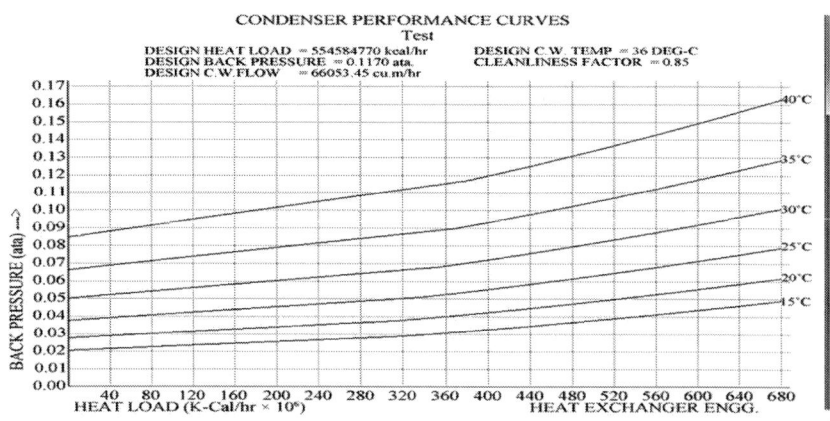

Figure 6: Condenser performance curve.

This parameter is required necessarily because it is the deciding factor for finalizing the size of the pump that is to be attached to the inlet water end. More is the Pressure Drop factor lesser capacity pump will be required and vice versa [12].

SOFTWARE EFFORT ESTIMATION

One of the major problems faced by project designers in controlling and managing software (SW) project is overrun of effort estimate [13]. The main objective of project designers is making correct decisions and leading to the development in relation to effect estimation. From the developed models, designers addressed that the effort estimate can be made earlier before the SW projects are fully developed, that means, the total Lines of Codes (LOC) of a SW project is counted only when the entire SW is completely developed. COCOMO is one well known method, to estimate the effort of SW projects automatically [14].

Effort Estimate Model

First, a nominal effort estimate is derived by Boehm using least square technique, which is a basic level of COCOMO. Secondly, the effort estimate is multiplied by a composite multiplier function, m(x), where x is the independent variables. The functional form of the model is defined as

$$\text{Effort Estimate} = E = A(i) \; S \; B(i) \; m(x)$$

where S is the size of LOC, m(x) is the multiplier function, and the coefficients A(i) and B(i) are derived from a combination of the mode and the level. In this SW A(i) and B(i) are 0.563 and 0.214. With respect to the size of LOC is 1467 and m(x) = 0.0078, then Effort estimate is 0.0209. Which is representing as less fault arises SW development.

Linear Regression Model (LRM)

Effort estimates are based on information available after detailed design or the program is fully developed [15]. The research model proposed with the following variables. The variable Effort (man-hours) spent by programmer to develop application SW. And other variables are:
  Design tools (DEGN_TOOL) = 2;

Programmer Experience (PROG_EXP) = 5;
Team Size (TEM_SIZE) = 2;
Program Complexity (PROG_COM) = 0.8;
Language Experience (LANG_EXP) = 10 yrs.

Regression Model Provides best Effort estimation by using Linear Regression Model. The LRM is hypothesized in the form of equation as

$$EFORT = A + B \times DEGN\_TOOL + C \times PROG\_EXP + D \times TEM\_SIZE + E \times PROG\_COM + F \times LANG\_EXP$$

where A, B, C, D, E, and F are to be determined by using normal equations.

Effort = 0.96 represent as high value effort.

Quality of Estimation

It indicates more accurate model, Magnitude of Relative Error (MRE) for estimate the quality of accuracy of the effort. The MRE is calculated by using the formula:

$$MRE = 100 \left| (Actual\ Effort - Estimated\ Effort) / Actual\ Effort \right|.$$

CONCLUSIONS

In this paper we have discussed a computer application that has been designed by keeping in mind the future enhancement of the work of power plants. The most lucrative feature of this application is that the three major modules designed have been given an independent bent of functionality. The advantage of this feature lies in the fact that as an effect of providing independency to each of the module, each on of them can be carried out as a Stand Alone Application. It actually helps different engineers to utilize specific modules according to their need. These further inches up the plausibility of development of separate modules of the project. In this way we have tried to develop a methodology that has completely computerized the designing activity of Condenser which used to be a manual activity amidst the development of power plant equipment—"Condenser".

REFERENCES

1. P. K. Giri and S. Srivastava, "Computer Application for Design Activity in Power Plants," International Conference on Modeling, Optimization, and Computing (ICMOS 2010), AIP Conference Proceedings, West Bengal, 28-30 October 2010, pp. 250-259. doi:10.1063/1.3516312
2. N. Hashemi, "A Relational Database Approach to Power Plant Design and Operating Plant Services," IEEE Transactions on Energy Conversion, Vol. 3, No. 3, 1988, pp. 487-490.
3. M. Botting and D. Kelly, "Computer Program Documentation System Condenser Design," David Taylor Model Basin Reports, DOME, 1966. http://oai.dtic.mil/oai/oai?verb=getRecord&metadataPrefix=html&identifier=AD0650571
4. K. S. S. Raj, "Deviations in Predicted Condenser Performance for Power Plants Using HEI Correction Factors: A Case Study," ASME Power Conference (POWER2006), Atlanta, 2-4 May 2006.
5. G. G. Rajan, "Energy Saving in Steam Systems". www.Klmtechgroup.Com.
6. J. Lang, "Design Procedure for Heat Exchangers on Aspen-Plus Software," Design Manual, 1999. http://cbe.sdsmt.edu/nsfproj/aspen/condenser.pdf
7. J. Lang, "Condenser Design on Aspen-Plus Software," ©SDSM&T, Rapid City, 2000.
8. Aspen Plus Simulator 10.0-1. User Interface, 1998.
9. http://pdf.ebooks6.com/Design-Procedure-for-a-Heat-Exchanger-on-the-AspenPlus-Software-pdf-e15431.pdf

10. X.-F. Dong, "Study on Intelligent Maintenance Decision Support System Using for Power Plant Equipment," 2008 Automation and Logistics, 2008. ICAL 2008. IEEE International Conference, Qingdao, 1-3 September 2008, pp. 96-100.
11. K. Darius, "Application of Intelligent Computer Aided Design Techniiques to Power Plant Design and Operation," Power Engineering Review, IEEE, PER-7, December 1987, pp. 34-35
12. S. I. K. Wu, "The Quality of Design Team Actors on Software Effort Estimation," Service Operations and Logistics, and Informatics (SOLI), IEEE International Conference, 21-23 June 2006.
13. S. Wu and I. Kuan, "A Component-Based Approach to Effort Estimation," The 4th International Conference on Wireless Communications, Networking and Mobile Computing (WiCOM 2008), Dalian, 12-14 October 2008, pp. 28-36.
14. J. C. Munson and T. M. Khoshgoftaar, "Regression Modeling of Software Quality: Empirical Investigation," Information and Software Technology, Vol. 32, No. 2, 1990, pp. 106-114. doi:10.1016/0950-5849(90)90109-5
15. P. K. Suri and P. Ranjan, "Comparative Analysis of Software Effort Estimation Techniques," International Journal of Computer Applications, Vol. 48, No. 21, 2012, pp. 12-19.

Chapter 3

A Novel Property of DNA – As a Bioflotation Reagent in Mineral Processing

Balasubramanian Vasanthakumar, Honnavar Ravishankar, and Sankaran Subramanian

Department of Materials Engineering, Indian Institute of Science, Bangalore, India

ABSTRACT

Environmental concerns regarding the use of certain chemicals in the froth flotation of minerals have led investigators to explore biological entities as potential substitutes for the reagents in vogue. Despite the fact that several microorganisms have been used for the separation of a variety of mineral systems, a detailed characterization of the biochemical molecules involved therein has not been reported so far. In this investigation, the selective flotation of sphalerite from a sphalerite-galena mineral mixture has been achieved using the cellular components of Bacillus species. The key constituent primarily responsible for the flotation of sphalerite has been identified as DNA, which functions as a bio-collector. Furthermore, using reconstitution studies, the obligatory need for the presence of non-DNA components as bio-depressants for galena has been demonstrated. A probable model involving these entities in the selective flotation of sphalerite from the mineral mixture has been discussed.

INTRODUCTION

The growing demand for mineral commodities across the world has led to the increased exploitation of lean grade ores with complex mineralogy, particularly for producing base metals. Additionally, depletion of the high grade mineral resources has resulted in the search for more advanced solutions to the problem of beneficiation of some refractory ores, in cases where conventional techniques are not efficient. These factors in combination with the more rigorous specifications for production of concentrates, stricter environmental legislation and a necessity to achieve lower operating costs has made it imperative to develop more effective flotation reagents.

Application of biotechnology in mineral processing has opened up immense possibilities for producing cleaner concentrates having acceptable grades and recoveries. Advances in molecular biology and genetic engineering have given an impetus for the characterization of biological entities of relevance to mineral processing. The bioflotation process may be defined as one in which microorganisms act as reagents, collectors or modifiers, to facilitate the selective separation of minerals [1]. The use of bioreagents as collectors invokes several interfacial aspects of the interacting biological and geological materials, viz., the physicochemical properties of the mineral surface, such as the atomic and electronic structure, the net charge/potential, the acid–base properties, and the wettability of the surface [2]. For the past two decades, several studies have been carried out on the use of microorganisms and their secretions viz., proteins and polysaccharides as environment friendly flotation reagents. The microorganism cell surface principally consists of functional groups derived from phospholipids, proteins and polysaccharides. Some of these induce hydrophobic properties, since they can adhere selectively to the mineral surface [3]. The emergence of mineral bioprocessing is reflected by the conferences exclusively dedicated to this topic [4], [5], [6], [7]. The bioflotation and bioflocculation processes of relevance to mineral beneficiation have been critically reviewed [1], [8]. These studies have led to the understanding that the interaction between specific microbial cells and mineral particles brings about significant changes in the chemistry of mineral surfaces, as well as the bacterial surfaces.

Despite the wealth of information that has been gathered so far, a detailed understanding of the nature and characteristics of the bioreagent responsible for the selective flotation of any given mineral from the mineral mixture has remained elusive. Such identification would constitute the first step in the large-scale generation and commercial exploitation of bioreagents involved in the flotation of minerals. In this communication, we report for the first time the identification and characterization of a bioreagent from B.circulans which plays a significant role in the selective flotation of sphalerite from a sphalerite-galena mixture. In addition, a plausible model for the interaction of this bioreagent with sphalerite leading to its selective flotation from the mineral mixture has been proposed.

RESULTS

Buffering of B.circulans Cells Aids the Flotation of Sphalerite

B.circulans suspended in water was tested for its capacity to float sphalerite or galena as a function of pH of the bacterial cell suspension. The lack of significant floatability of either sphalerite or galena throughout the pH range tested, prompted the exploration of the circumstances under which the bacteria could float the minerals (Fig.1). Since most biological reactions operate under buffered conditions [9], the role of anionic and cationic buffer systems was evaluated in the flotation process. Curiously, only cells suspended in anionic buffers promoted the flotation of sphalerite while cationic buffers did not aid the flotation process. The maximum flotation of sphalerite was observed with phosphate buffer (pH 8.0). Fig. 1 shows a representative trend obtained with an anionic buffer (phosphate) and a cationic buffer (tris-HCl). The intriguing and antagonistic role of the anionic and cationic buffers is presumably related to the nature of the bioreagent (see Discussion). Based on these observations, all the subsequent flotation studies were carried out in the presence of bacterial cells suspended in phosphate buffer (pH 8.0). Buffers per se failed to enhance the flotation recovery of minerals beyond background levels (5%). In the case of galena, neither anionic nor cationic buffers could aid its flotation. These observations indicate that the presence of a component on the surface of B.circulans cells which aids the flotation of sphalerite but not galena.

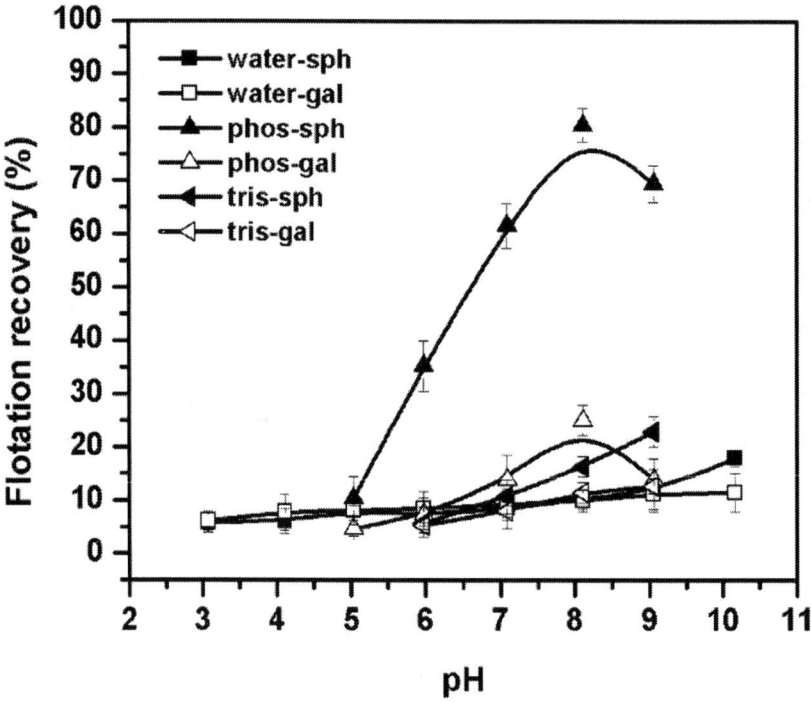

Figure 1: Effect of buffering of cells on the flotation recovery of sphalerite and galena.

Nature of the Bioreagent Aiding Sphalerite Flotation

The cell wall architecture of Gram positive bacteria such as Bacillus is made up of membrane proteins and a thick peptidoglycan layer connected by wall teichoic acids [10]. Encouraged by the high flotation levels of sphalerite but not galena in the above experiments, an enzymatic method was adopted to identify the nature of the bioreagent(s) involved in the above process.

To this end, live cells were treated with enzymes to digest away one surface component at a time. The effect of its presence or absence on the flotation of sphalerite was then evaluated. It is pertinent to mention that the enzymatic treatments did not affect the viability of the cells. As shown in Fig. 2, treatment of cells with either proteinase K or lysozyme did not reduce the flotation recovery of sphalerite significantly compared with the untreated control cells. This suggested that the cell surface component aiding flotation is non-proteinaceous and non-polysaccharide in nature. Teichoic acid, which is another major component of cell walls could not be assayed as above, as no enzymatic activity degrading teichoic acid was available commercially. As phosphate is one of the principal constituents of teichoic acids [11] an indirect method of culturing B.circulans cells in a phosphate deficient medium was resorted to address the question of its involvement in the flotation process. Cells grown in such a deficient medium showed considerably reduced growth and low teichoic acid content, which are in agreement with similar observations reported earlier [12]. The flotation recovery of sphalerite by these cells was reduced to about 30%, but not inhibited completely (Fig.2). This result suggested that, plausibly, some phosphate-dependent component was aiding flotation. If teichoic acid indeed were to be aiding flotation, purified teichoic acid would be expected to show high flotation recovery of sphalerite, similar to normal cells. But when purified teichoic acid was used in the flotation assay, the recovery of sphalerite was very low (~10%). This confirms that teichoic acids were not involved in the flotation of sphalerite.

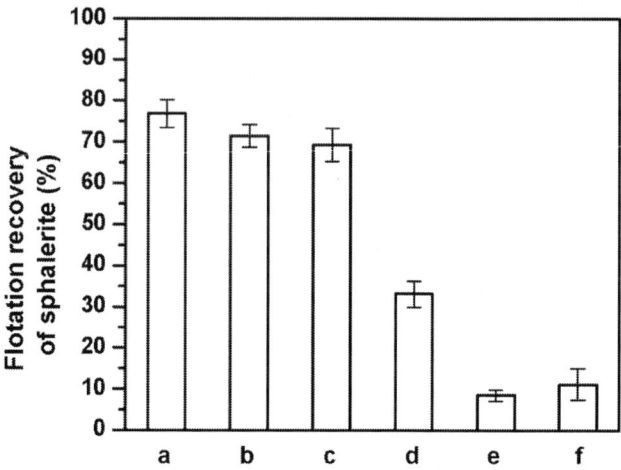

Figure 2: Flotation recovery of sphalerite in the presence of bacterial cells, before and after various enzymatic treatments. a - untreated cells; b - proteinase K treated; c - lysozyme treated; d - phosphate deficient medium; e - purified teichoic acid; f – DNase 1 treated.

A major phosphate-dependent constituent of cells is nucleic acids. Several bacterial species are known to secrete nucleic acids on to their cell surface. This double stranded DNA has been designated as extracellular DNA (eDNA) to distinguish it from genomic DNA, which is intracellular in location [13]. Surprisingly, this aspect of bacterial cell surface has not received sufficient attention. In contrast to the above enzymatic treatments, it is noteworthy that DNase 1 treatment of cells markedly reduced the flotation recovery of sphalerite to 10%. These experiments clearly indicate that the bioreagent responsible for sphalerite flotation is a non-teichoic acid entity, which is proteinase K and lysozyme resistant, but DNase 1 sensitive.

Selective Flotation of Sphalerite from a Sphalerite – Galena Mineral Mixture by Thermolysed Cells and its Fractionated Components

In the above context, for any reagent to be considered promising for further development, it should be able to selectively float the mineral of interest from a mineral mixture. For this study, a 1:1 mixture of sphalerite and galena was chosen and the selectivity index was calculated based on equation (1). As shown in Fig.3 normal cells are able to partially float sphalerite from the mineral mixture with a selectivity index of 5.5. In this case, sphalerite and galena were floated to the extent of 28% and of 3% respectively. This recovery is much lower than that observed with the flotation of sphalerite by cells, where almost 80% of the mineral was floated (Fig.1). The difference between the two observations could be due to the high level of unproductive adsorption of cells on to galena vis-à-vis sphalerite [14]. In order to enhance the selective recovery of sphalerite it was evaluated whether viability or intactness of B.circulanscells is necessary to float sphalerite. If the observations of the previous section were to be true, thermolysed cells should show higher selective flotation recovery of sphalerite from the mineral mixture compared with live cells. Thermolysis results in rupturing of the bacterial cell structure and releasing all the intracellular contents into the medium [15]. One of the chief components of interest, vis-à-vis sphalerite flotation, which is released into the medium is genomic DNA. Thus, higher flotation recovery of sphalerite is expected to be due to the combined effect of the intracellular and the extracellular DNA that results from cell disruption.

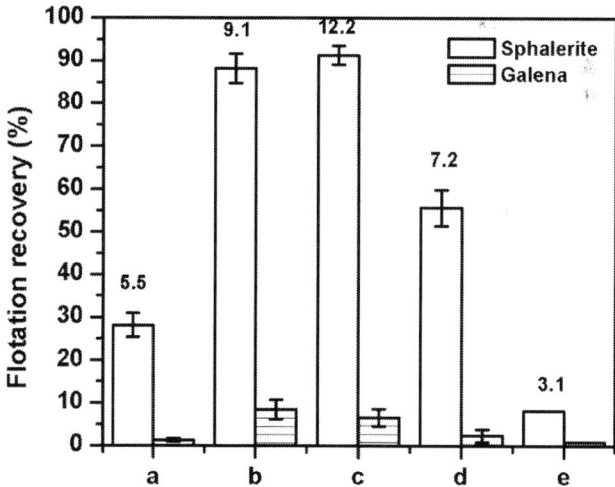

Figure 3: Selective flotation of sphalerite and galena in the presence of unfractionated or thermolysed and fractionated components of B.circulans. a

- normal cells; b - thermolysed cells; c - thermolysed cell-free supernatant; d - thermolysed cell pellet; e – DNase 1 treated thermolysed cell-free supernatant. The numbers above the bar chart indicates the selectivity index (S.I) values.

As shown in the Fig. 3 this expectation was borne true. Interaction with the thermolysed cells resulted in the flotation recovery of 90% for sphalerite and 10% for galena to give a selectivity index of 9.1. Thermolysed cells were subsequently fractionated and assayed for the selective flotation of sphalerite. This was carried out to determine the extent of partitioning of the bioreagent into the supernatant and pellet fractions. The thermolysed cell-free supernatant behaved slightly better than the thermolysed cells in that 92% of sphalerite and 7% of galena were floated to give a selectivity index of 12.2. The cell pellet fraction, however, yielded a flotation recovery of 55% of sphalerite and 5% of galena with a lower selectivity index of 7.2. Furthermore, DNase 1 treatment of the thermolysed cell-free supernatant significantly reduces the selective flotation of sphalerite, again indicating that DNA aids sphalerite flotation. This experiment clearly demonstrates that B.circulans cells or its constituents selectively float sphalerite from a sphalerite-galena mixture. Disruption of cell structure by thermolysis enormously enhances the flotation recovery of sphalerite from the mineral mixture and a majority of the entity aiding sphalerite flotation is released into the supernatant.

Purified DNA Floats Sphalerite but not Galena

In order to conclusively prove that DNA is the entity aiding sphalerite flotation, genomic DNA purified from B.circulans cells was used for flotation tests. As shown in Fig. 4 using purified genomic DNA as collector (2 mg dsDNA which is the equivalent of that present in thermolysed cell-free supernatant), the flotation recovery of about 50% of sphalerite could be achieved. Treatment of genomic DNA with DNase 1 prior to interaction with sphalerite, significantly diminishes its flotation (<10%). However, using the same amount of thermolysed genomic DNA (ssDNA), greater than 80% recovery of sphalerite could be achieved. This is comparable to that obtained with thermolysed cell-free supernatant. The conversion of dsDNA to ssDNA leads to a effective doubling of the bio-collector concentration. As observed previously, treatment of the thermolysed genomic DNA with DNase 1 prior to interaction with sphalerite, almost completely retards its flotation (<10%). This experiment clearly establishes that the reagent involved in the flotation process in thermolysed cells or its supernatant is ssDNA and the lower flotation recovery obtained with viable cells is due to the dsDNA present on its cell surface. Studies carried out by others regarding the source of this eDNA on bacterial surfaces have indicated that it arises from the intercellular genomic DNA via multiple ways [16]. This experiment also establishes that ssDNA acts as high capacity bio-collector compared to equivalent amounts of dsDNA in the flotation of sphalerite.

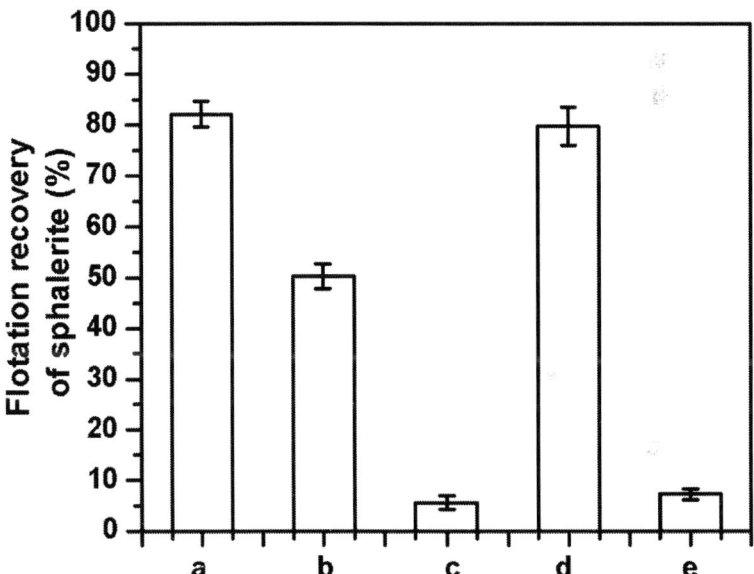

Figure 4: Effect of the strandedness of DNA on the flotation of sphalerite. a - thermolysed cell-free supernatant; b - genomic DNA (2 mg of dsDNA); c – DNase 1 treated genomic DNA; d - ssDNA (2 mg); e - DNase 1 treated ssDNA.

Fig. 5 shows the concentration dependent recovery of sphalerite and galena by purified ssDNA in a flotation experiment. The flotation recovery of sphalerite reaches a maximum (85%) in the presence of 2 mg of ssDNA, beyond which it remains unchanged. The flotation of galena also shows a steady marginal increase up to about 2 mg of ssDNA and then attains a plateau. However, the maximum flotation of galena observed is 30%. These profiles reflect largely the relative recoveries of sphalerite and galena observed with intact cells or its components.

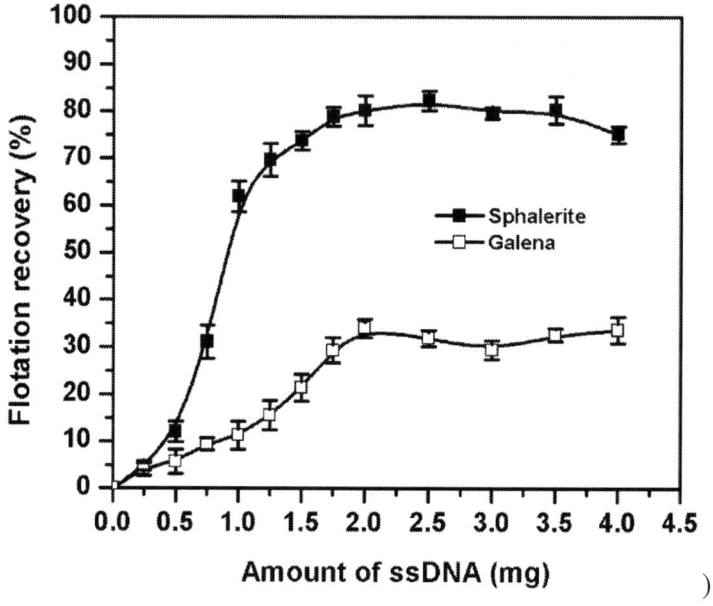

Figure 5: Flotation recovery of sphalerite and galena as a function of the ssDNA concentration.

Amphipathic Nature of ssDNA Facilitates Sphalerite Flotation

It thus becomes of interest to understand the mechanism of ssDNA induced flotation of sphalerite. An examination of the structure of DNA indicates its amphipathic nature with the polyphosphate groups aligned on one face of the long axis and the hydrophobic bases aligned on the opposite face (Fig. 6a). At pH 8.0, DNA behaves as a polyanionic species [17]. In the case of dsDNA, the negatively charged phosphates are present on the outside of the double stranded helix while the stacked hydrophobic bases are buried on the inside of the helix due to base pairing of the two strands. Thus pairing of strands evidently reduces its amphipathic nature. In contrast to this, the unpaired state of ssDNA makes its amphipathic nature more conspicuous. The higher intrinsic amphipathic nature, along with an increase in the effective bio-collector concentration may be responsible for the higher flotation recovery of sphalerite by ssDNA vis-à-vis dsDNA (Fig. 4). We hypothesized that the above structure could lead to strong interaction between sphalerite and ssDNA such that the phosphate groups could interact with the sphalerite while the aromatic bases could align across the axis to form the hydrophobic surface. Thus ssDNA could act as a polymeric heteropolar collector. To validate this hypothesis we evaluated the involvement of both the surfaces of this amphipathic molecule in the flotation process.

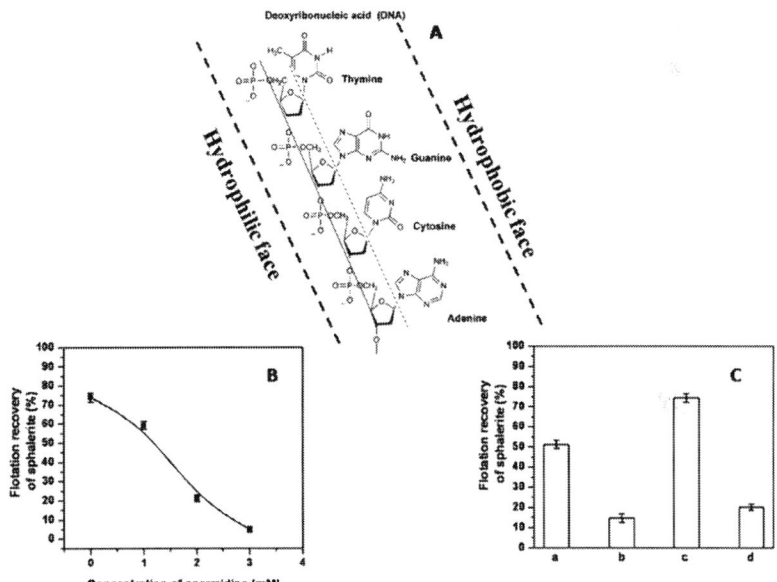

Figure 6: Schematic diagram of the amphipathic nature of ssDNA. (A) Amphipathic structure of ssDNA indicating the hydrophobic and the hydrophilic faces; (B) - Effect of spermidine concentration on the ssDNA mediated flotation of sphalerite; (C) - Effect of depurination of DNA on the flotation of sphalerite a - dsDNA; b - depurinated dsDNA; c - ssDNA; d - depurinated ssDNA.

The hydrophilic face containing the phosphate groups is not amenable to removal, as it constitutes the backbone of the DNA polymer.

However, the effect of the hydrophilic face of the DNA molecule can be reduced by the electrostatic interactions of phosphate groups with inorganic and other small cationic molecules in a concentration dependent manner [18], [19]. The neutralization of the polyanionic character of ssDNA with spermidine (a cellular cationic small molecule) to abolish or interfere in the flotation recovery of sphalerite was evaluated. Essentially, there is a competition between the mineral and spermidine for the polyanionic ssDNA. It was ensured that the concentration of spermidine used in this experiment (upto 3 mM) is several fold lower than that known to precipitate DNA (>10 mM) [20]. As shown in Fig 6b, with increasing concentration of spermidine the flotation of sphalerite continuously decreases. This attests to the need for the charged phosphate groups in ssDNA for effective flotation to take place.

While the polyanionic ssDNA interacts with the mineral, through chemical and electrostatic forces, the aromatic bases present on the outer surface presumably provide the necessary hydrophobicity for the flotation process. To ascertain if this indeed were to be the case, a considerable fraction of the bases was removed by depurination before interacting with the mineral. An otherwise continuous stretch of hydrophobic face of DNA appears discontinuous and punctured with patches of hydrophilic holes following acid depurination. If hydrophobicity of DNA is indeed necessary for the flotation of sphalerite, any reduction in this property would negatively impact the flotation process. The effect of depurination of dsDNA and ssDNA on the flotation of sphalerite was assessed. Previous estimations had indicated that close to 50% of the bases are converted to apurinic acid by this treatment [21]. As anticipated, the flotation recoveries were drastically reduced after depurination of either dsDNA or ssDNA (Fig.6c). This clearly indicated that the aromatic entities of ssDNA imparted the necessary hydrophobicity for the flotation process. The foregoing data demonstrates that the amphipathic nature of ssDNA is absolutely essential for the successful flotation of sphalerite.

Non-DNA Components of Lysed Cells Act as a Depressant for Galena in the Selective Flotation Process

Having identified DNA, or more specifically ssDNA, as the bio-collector responsible for the high flotation recoveries of sphalerite, it becomes of interest to reconstitute the process of selective flotation. It has been observed that ssDNA could float sphalerite but not galena. Surprisingly, in contrast to the thermolysed cell-free supernatant, ssDNA per se showed no selectivity towards sphalerite in the sphalerite-galena mixture. In fact, hardly any flotation of either mineral was observed (Fig.7). This indicated that other (non-DNA) components of the thermolysed supernatant may also be involved in the selective flotation of sphalerite. However, by adding varying amounts of DNase 1 treated thermolysed cell-free supernatant to a defined amount of ssDNA prior to interaction with the mineral, the degree of selective flotation of sphalerite from the sphalerite-galena mixture could be restored to that observed in the case of sphalerite when present alone. Importantly, DNase 1 treated thermolysed cell-free supernatant by itself did not selectively float sphalerite (refer Fig.3). This experiment clearly reveals that ssDNA which can float sphalerite is ineffective for the selective flotation of sphalerite from the sphalerite-galena mixture. It needs other non-DNA components of the thermolysed cell-free supernatant, bulk of which appears to be polyanionic, to selectively float sphalerite from its mixture with galena. These non-DNA polyanions presumably compete with ssDNA for adsorption on to the mineral surface and prevent the non-specific and non-productive adsorption of ssDNA on galena. These experiments clearly demonstrate the separation of bio-collector and bio-depressant functions between ssDNA and the non-DNA components of thermolysed cells. Reconstitution data presented in Table - 1 clearly shows that the selectivity index values with other Bacillusspecies used in the study follow a similar trend during the selective flotation of sphalerite from the mineral mixture.

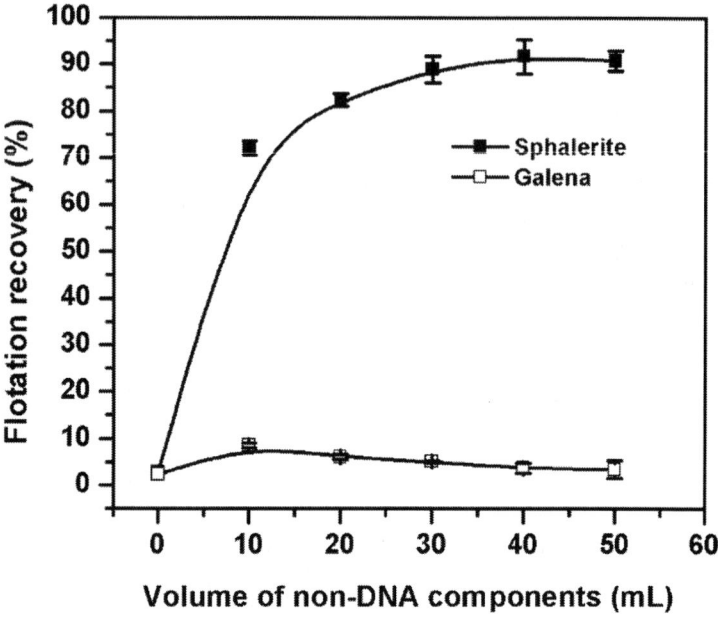

Figure 7: Flotation recovery of sphalerite and galena during reconstitution of the selective flotation of sphalerite with a fixed amount (2 mg) of ssDNA and varying amounts of DNase 1 treated non-DNA components of cells.

Table 1: Organism dependent Selectivity Index values during reconstitution of selective flotation tests using fixed amount (2 mg) of ssDNA

Volume of non-DNA components (mL)	B.circulans	B.megaterium	B.subtilis	P.polymyxa
0	1.1	1.7	1.3	1.4
10	5.3	3.1	2.7	3.6
20	8.5	5.6	2.9	5.9
30	12.2	8.3	5.1	8.6
40	16.5	15.3	9.4	10.8
50	16.9	19.1	9.9	12.0

doi:10.1371/journal.pone.0039316.t001

DISCUSSION

Traditionally, DNA has always been ascribed biological roles which are central to all life forms. Of late, some non-biological roles such as its use in nano-technology has been recognized [22],[23]. In this communication, we have identified that DNA, or more specifically ssDNA, acts as bio-collector in the selective floatation of sphalerite from a sphalerite-galena mineral mixture. This indeed is a novel property of DNA and one that is very different from the biological properties of DNA identified so for. We have demonstrated that the amphipathic nature of ssDNA facilitates the flotation process.

Earlier studies from this laboratory have shown that the capacity of galena to adsorb P. polymyxa cells is an order of magnitude higher than that of sphalerite [14]. This situation likely appears to be true for B.circulans also. When pure ssDNA is used in the selective flotation of sphalerite from a sphalerite-galena mixture, the flotation recovery is found to be highly reduced compared to that with sphalerite alone. Addition of several fold excess of ssDNA does not significantly enhance the flotation recoveries of sphalerite from the mixture, presumably due to the higher preferential adsorption of ssDNA by galena over sphalerite. However, in the presence of a large excess of other polyanionic species in the thermolysed cell-free supernatant (which comprises of all the cellular non-DNA components), ssDNA is largely left free to bind with sphalerite and aid its flotation, while the non-DNA components preferentially binds with galena. In this context, polysaccharides have been shown to act as a depressant in the flotation of sulfide minerals [24], [25]. Reconstitution studies presented above indicate that the ratio of the bio-collector and bio-depressant probably determines the flotation recoveries of sphalerite from the mineral mixture.

Based on the above investigations, we present a model for the flotation of sphalerite from a sphalerite-galena mixture. This process needs three essential components.

- an anionic buffer component
- a bio-collector (ssDNA) which facilitates the flotation process
- bio-depressants which retard the flotation process. This comprises of teichoic acids and polysaccharides, which are essentially non-amphipathic polyanions.

The antagonistic role of anionic and cationic buffers, rather than its pH, in the flotation of sphalerite calls for an explanation. A probable reason for the negative effect of cationic buffer on the flotation process may be due to the competition of the mineral surface with metal cations for interaction with the poly-anionic ssDNA. The buffer cations being far in excess, probably out compete the metals to bind ssDNA. In contrast to this, an anionic buffer such as phosphate compete with the poly-anionic ssDNA for the metal surface. The poly-anionic nature of ssDNA compared to the monomeric nature of the buffer anion leads to a synergistic interaction with the metal cations on the mineral surface. Thus, the presence of relevant factors that lead to better mineral-ssDNA interaction and the resultant induction of hydrophobicity on the mineral surface has consequences for high flotation recovery.

Metal cations have been shown to have two modes of interaction with the constituents of DNA, viz., phosphates and bases. Pb is known to interact more avidly with both the phosphate backbone and aromatic bases than Zn [26]. Additionally, the lattice structure of sphalerite rather than galena may be more amenable to a favorable interaction with ssDNA. For these reasons, ssDNA may not be able to induce sufficient hydrophobicity to galena, leading to a low recovery.

In summary, a novel bio-collector property of ssDNA for sphalerite flotation has been demonstrated. Furthermore, the twin presence of the bio-collector and bio-depressants is absolutely essential to achieve higher selective flotation of sphalerite from a sphalerite-galena mixture.

MATERIALS AND METHODS

Mineral samples of sphalerite and galena were obtained from Wards Natural Science Establishment (USA) and Alminrock Indscer Fabriks (India) respectively. Mineralogical studies as well as X-ray powder diffraction data indicated that the samples were of high purity (99.8%). The above samples were dry ground using a porcelain ball mill and then sieved through BSS sieves.

B. circulans (NCIM 2160), B. subtilis (NCIM 2063), B. megaterium (NCIM 2087), P. polymyxa(NCIM 2539) and E. coli K12 (NCIM 2674) used in this study were obtained from the National Collection of Industrial Microorganisms (NCIM), National Chemical Laboratory, Pune. The bacteria were cultured using the Bromfield medium as described elsewhere [27]. Phosphate deficient Bromfield medium had a phosphate content of 0.05 g/L. Spermidine was obtained from Sigma Aldrich while the enzymes used in this study, viz., proteinase K, lysozyme and DNase 1 were obtained from Bangalore Genei.

Harvesting, Cell Fractionation and Extraction of Genomic DNA

Cells were harvested from fully grown bacterial cultures (48 h) by centrifugation at 5000 rpm for 20 min at 4°C. Bacterial cells (1×10^{10}) washed and suspended in 0.1 M phosphate buffer (pH 8.0) were used directly or thermolysed in a water bath at 100°C for 30 min and cooled before being used in the flotation tests. The thermolysed cell suspension was centrifuged at 10,000 rpm for 20 min at 4°C to obtain the soluble thermolysed cell-free supernatant and insoluble thermolysed cell pellet. The thermolysed cell-free supernatant was assayed directly while the insoluble pellet was suspended in 0.1 M phosphate buffer before assaying.

Extraction of genomic DNA was essentially carried out as per the procedure described elsewhere [28]. Tris HCl (pH 8.0) was however replaced by 0.1 M phosphate buffer (pH 8.0). In this communication, genomic DNA, which is double stranded, has been referred to interchangeably as dsDNA and thermolysed genomic DNA, which is single stranded, as ssDNA.

Enzymatic Treatment of Cells

Bacterial cells (1×10^{10}) suspended in 0.1 M phosphate buffer (pH8.0) were treated with enzymes at the indicated concentrations (1 mg/mL proteinase K, 10 mg/mL lysozyme or 200 units DNase 1 in 10 mM $MgCl_2$) separately for 4 h. After treatment, the cells were centrifuged at 5000 rpm for 20 min, washed and suspended in 0.1 M phosphate buffer before being used in the flotation experiments.

Extraction of Teichoic Acid from Cells

This was carried out as described elsewhere [29]. Purified teichoic acid was dissolved in 0.1 M phosphate buffer before using it in the flotation tests of sphalerite or galena.

Flotation of Sphalerite or Galena and Selective Flotation of Sphalerite from a Mixture of Sphalerite and Galena

1 g of sphalerite or galena of size (−150+100) µm was used for the flotation experiments. For selective flotation a mixture of sphalerite and galena of size (−150+100) µm in the ratio of 1:1 (0.5 g each) was used. Pure mineral or the mineral mixture was conditioned with the chosen reagent (whole cells/thermolysed cells/thermolysed cell-free supernatant/thermolysed cell pellet) in 0.1 M phosphate buffer at pH 8.0 for 30 minutes prior to the flotation process. After interaction the suspension was transferred to a modified Hallimond tube [30]. Nitrogen gas at a flow rate of 40 mL/min was passed through the cell and the flotation was carried out for 3 minutes. The concentrate and tailing fractions were separately filtered, dried and weighed. Lead and zinc contents in the concentrate and tailing fractions were estimated using an Atomic Absorption Spectrometer (AAS, Thermo Electron Corporation MM series). Selectivity Index was calculated according to Gaudin's formula [31] as shown in equation (1).

$$S.I = \sqrt{\frac{R_a \times J_b}{(100 - R_a) \times (100 - J_b)}} \tag{1}$$

where, R_a – Percentage recovery of sphalerite in the float fraction, J_b – Percentage recovery of galena in the tailings fraction.

Depurination of DNA and Interaction of ssDNA with Spermidine

Depurination of DNA was carried out as described [21]. Genomic DNA (2 mg) was adjusted to pH 1.6 with HCl in the presence of 50 mM NaCl and 1.5 mM of sodium citrate and dialyzed for 15 hours at 37°C. The dialysate was either thermolysed or used directly for the flotation of sphalerite.

ssDNA obtained from thermolysis of 2 mg of dsDNA in 0.1 M phosphate buffer (pH 8.0) was interacted with varying concentrations of spermidine for 2 h at 37°C as described [32] prior to its use in the flotation of sphalerite.

Reconstitution of the Selective Flotation Process

Selective flotation was reconstituted by adding ssDNA (obtained from 2 mg of thermolysed genomic DNA) or ssDNA together with varying amounts of DNase 1 treated thermolysed cell-free supernatant to the mineral mixture. The flotation procedure as described earlier was adopted.

All the above experiments were carried out atleast three times to determine the standard error.

AUTHOR CONTRIBUTIONS

Conceived and designed the experiments: SS RS. Performed the experiments: BV. Analyzed the data: SS RS BV. Wrote the paper: SS RS.

REFERENCES

1. Rao KH, Subramanian S (2007) in Bioflotation and Bioflocculation of relevance to minerals bioprocessing, Donati ER, and Sand W editors. Microbial Processing of Metal sulfides, (Springer), pp 267–286.
2. Rao KH, Vilinska A, Chernyshova IV (2010) Minerals bioprocessing: R & D needs in mineral biobeneficiation. Hydrometallurgy 104(3–4): 465–470.
3. Smith RW, Miettinen M (2006) Microorganisms in flotation and flocculation: Future technology or laboratory curiosity? Minerals Engineering 19(6–8): 548–553.
4. Smith RW, Misra M editors (1991) Mineral Bioprocessing. (The Minerals Metals and Materials Society, PA).

5. Holmes DS, Smith RW editors (1995) Mineral Bioprocessing vol II. (The Minerals Metals and Materials Society, PA).
6. Kuyucak N editors (1998) Third International Conference on Minerals Bioprocessing and Biorecovery/Bioremediation in Mining. (Gordon and Breach Science).
7. Rao KH, Forssberg KSE editors (2001) Mineral Bioprocessing IV Int J Miner Process vol 62 Special issue.
8. Chandraprabha MN, Natarajan KA (2009) Microbially induced mineral beneficiation. Min. Proc. Ext. Met. Review 31: 1–29.
9. Stoll VS, Blanchard JS (1990) Buffers: Principles and practice. Methods Enzymol 182: 24–38.
10. Vollmer W (2012) Bacterial outer membrane evolution via sporulation? Nat Chem Biol 8: 14–18.
11. Magda LA, Pereira PM, Yates J, Reed P, Veiga H, et al. (2010) Teichoic acids are temporal and spatial regulators of peptidoglycan cross-linking in Staphylococcus aureus. Proc Natl Acad Sci USA 107 (44): 18991–18996.
12. Bhavsar AP, Erdman LK, Schertzer JW, Brown ED (2004) Teichoic acid is an essential polymer in Bacillus subtilis that is functionally distinct from teichuronic acid. J. Bacteriol. 186: 7865–7873.
13. Draghi JA, Turner PE (2006) DNA secretion and gene-level selection in bacteria. Microbiology 152: 2683–2688.
14. Santhiya D, Subramanian S, Natarajan KA (2001) Surface chemical studies on sphalerite and galena using Bacillus polymyxa: I. Microbially induced mineral separation. J Colloid Interface Sci. 235(2): 289–297.
15. Harrison STL (1991) Bacterial cell disruption: A key unit operation in the recovery of intracellular products. Biotech. Adv. 9(2): 217–240.
16. Flemming H-C, Wingender J (2010) The biofilm matrix. Nat Rev Microbiol. 8: 623–633.
17. Perks B, Shute JK (2000) DNA and Actin Bind and Inhibit Interleukin-8 Function in Cystic Fibrosis Sputa: In Vitro Effects of Mucolytics. Am. J. Respir. Crit. Care Med. 162: 1767–1772.
18. Felsenfeld G, Miles HT (1967) The physical and chemical properties of nucleic acids. Annu Rev Biochem 36: 407–448.

19. Izatt RM, Christensen JJ, Rytting JH (1971) Sites and thermodynamic quantities associated with proton and metal ion interaction with ribonucleic acid, deoxyribonucleic acid, and their constituent bases, nucleosides, and nucleotides. Chem. Rev. 71(5): 439–481.
20. Pelta J, Livolant F, Sikorav J-L (1996) Nucleic acids, protein synthesis, and molecular genetics. J. Biol. Chem. 271: 5656–5662.
21. Huang PC, Rosenberg E (1966) Determination of DNA base composition via depurination. Anal Biochem. 16(1): 107–113.
22. Xia F, Zuo X, Yang R, Xiao Y, Kang D, et al. (2010) On the Binding of Cationic, Water-Soluble Conjugated Polymers to DNA: Electrostatic and hydrophobic interactions. J. Am. Chem. Soc. 132(4): 1252–1254.
23. Wickham SFJ, Bath J, Katsuda Y, Endo M, Hidaka K, et al. (2012) A DNA-based molecular motor that can navigate a network of tracks. Nat. Nanotech. doi 10.1038/nnano.2011.253.
24. Rath RK, Subramanian S (1999) Adsorption, electrokinetic and differential flotation studies on sphalerite and galena using dextrin. Int. J. Min. Process. 57(4): 265–283.
25. Santhiya D, Subramanian S, Natarajan KA (2002) Surface chemical studies on sphalerite and galena using extracellular polysaccharides isolated from Bacillus polymyxa. J Colloid Interface Sci. 256(2): 237–248.
26. Duguid J, Bloomfield VA, Benevides J, Thomas Jr GJ (1993) Raman spectroscopy of DNA-metal complexes. I. Interactions and conformational effects of the divalent cations: Mg, Ca, Sr, Ba, Mn, Co, Ni, Cu, Pd, and Cd. Biophys. Journal 65(5): 1916–1928.
27. Bromfield SM (1954) Reduction of ferric compounds by soil bacteria. J Gen Microbiol. 11: 1–6.
28. Sambrook J, Fritsch EF, Maniatis T editors (1989) Molecular cloning. A laboratory manual. 2nd edition. (Cold Spring Harbor Laboratory Press, NY).
29. Räsänen L, Mustikkamäki UP, Arvilommi H (1982) Polyclonal response of human lymphocytes to bacterial cell walls, peptidoglycans and teichoic acids. Immunology 46(3): 481–6.
30. Fuerstenau DW, Metzger PH, Seele GD (1957) Modified Hallimond tube for flotation testing. Eng. and Mining Journal 158: 93–95.

31. Gaudin AM (1957) Flotation 2nd edition (McGraw-Hill, New York) pp 573.
32. Rubin RL (1997) Spermidine-Deoxyribonucleic acid interaction in vitro and inEscherichia coli. J. Bacteriol. 129: 916–925.

Chapter 4

Acoustical Design of an Electrical Emergency Plant Using Sea Method

Evgeny Podzharov[1], José F. de la Mora Gálvez[2], and Jesus A. Alvarez Sanchez[1]

[1]Electromechanical Engineering Department, University of Guadalajara, Puerta 10, Guadalajara, México

[2]Superior School of Engineering Prolongacion Calz, University of Panamericana, Circunv. Pte 49, Guadalajara, México

ABSTRACT

The statistical energy analysis (SEA) was used in the acoustical design of an electrical emergency plant to reduce the outdoor noise level. In the past, when the plant was working, a high annoying noise was heard all over the university camp. At a first glance the principal ways of noise propagation were the open door of the plant which was used for the suction of fresh air and a vast hole in the ceiling which was used for gases outlet. Also, a spectral analysis of the noise inside the plant showed that the dominant frequencies of the noise were in the range of 120 - 270 Hz. This frequency range is near the critical frequency of the brick walls that is 129 Hz, at which the walls are transparent for noise. A two-block diagram is used for the statistical energy analysis. Two ways of sound transmission are considered through the inlet and outlet holes and through the walls and ceiling. This analysis shows that the exclusion of holes wouldn't be sufficient to reduce noise to an acceptable level in a low frequency range but increase the noise absorption by the wall coating material. The transmission loss is calculated for different wall coatings and hole areas. A layer of fiberglass of two-inch width is selected to increase the wall absorption coefficient. Special silencers are designed and put at the suction of air and at the outlet of engine gases to reduce the noise propagation through the holes. The noise measurement shows that the noise level is considerably reduced after implementation of these measures. The reduction of noise is 7 - 8 dB (A), 19 dB (A) and 23 dB (A), inside the plant, 10 m and 15 m away from the plant, respectively.

INTRODUCTION

The noise contamination in the cities is an important problem nowadays as the cities become more populated and there are more vehicles on the streets. One part of the problem is the noise in studying and working areas as well as in the residential areas.

In the University of Panamericana in Guadalajara, Mexico, there is an electrical emergency plant which supplies electricity for the campus during the electricity cut-offs in the summer during the rainy season. The plant has a diesel engine and electric generator which are installed inside a small building. This engine was producing a very loud noise which was heard all over the university campus and was the cause of many complaints by students and professors.

NOISE MEASUREMENTS AND ITS ANALYSIS

The results of the noise measurements inside and outside of the plant are shown in Figure 1. The noise level inside the plant is very high (112 dBA). It decreases 13 dBA passing the open door of the plant. Then it decreases slowly till 72 dBA at 25 m from the source being still high. The measurements in the offices and the class rooms of the building nearer to the plant showed noise levels from 59 dBA to 69 dBA which are all beyond the permissible noise level.

The noise spectrum inside the plant shows (Figure 2) that the majority of noise's components are in low frequency range near 200 Hz; meanwhile there are also high level components in the middle frequency range up to 1200 Hz and in higher frequencies.

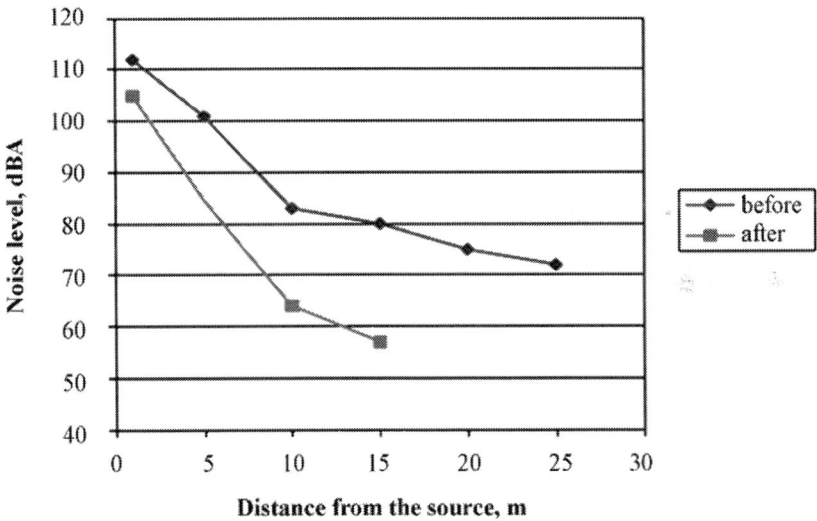

Figure 1: Noise level inside and outside of the plant.

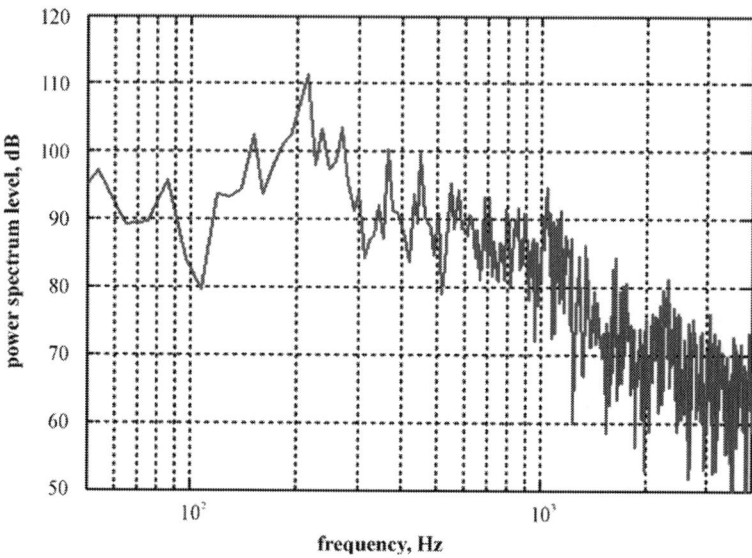

Figure 2: Noise spectrum measured inside of the plant.

ANALYSIS OF THE PLANT ACOUSTICAL SYSTEM USING SEA METHOD

The principle ways of noise propagation must be studied to reduce effectively the noise level. The noise radiated by the machine towards the internal acoustical space inside the plant goes out to the external acoustical space by three ways: 1) through the entrance door and the exhaust hole in the ceiling, 2) through the walls, 3) through the foundation. The third way can be considered insignificant because the machine is properly vibroisolated from the foundation.

To realize the acoustical analysis of the plant by the SEA method an acoustical model in Figure 3can be used.

Where E1—acoustical energy of the internal acoustical space, J E2—acoustical energy of the external acoustical space, J W in—acoustical power introduced by the machine, W

W_i^d—acoustical power dissipated in the element i of the model, which can be determined using the following equation [1]

$$W_i^d = \eta_i E_i \omega,\qquad(1)$$

Where

η_i—loss factor of element i $\omega = 2\pi f$—angular frequency, rad/s f—frequency, Hz.

W'_{12}—acoustical power transmitted through the walls and ceiling, W

W''_{12}—acoustical power transmitted through the door and the air duct in the ceiling, W According to [1] [2] the equation of energy balance can be written as follows

$$W_1^{in} = W_1^d + W'_{12} + W''_{12},$$

$$W_2^d = W'_{12} + W''_{12}.\qquad(2)$$

Here

$$W'_{12} = \omega(\eta'_{12} E_1 - \eta'_{21} E_2),$$
$$W''_{12} = \omega(\eta''_{12} E_1 - \eta''_{21} E_2)\qquad(3)$$

Where

η_{ij}—transmission loss factor between elements i and j.

Substituting Equation (3) into Equation (2) and then into Equation (1) and considering that acoustical energy transmission is insignificant in the inverse direction, as $\eta_{12} \gg \eta_{21}$, the equation of energy balance can be obtained in the following form

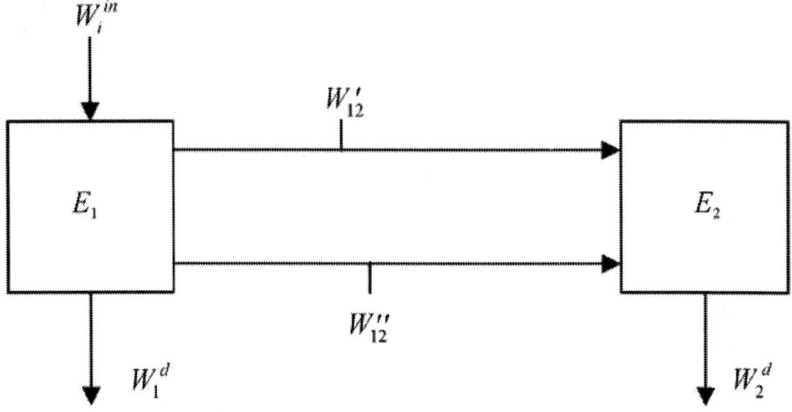

Figure 3: Acoustical model of the plant.

$$W_1^{in} = E_1\omega(\eta_1 + \eta'_{12} + \eta''_{12}), \tag{4}$$

$$W_2^d = E_1\omega(\eta'_{12} + \eta''_{12})$$

Now, dividing the first equation of (4) by the second, we have

$$W_1^{in}/W_2^d = 1 + \eta_1/(\eta'_{12} + \eta''_{12}) \tag{5}$$

The acoustical energy transmitted through the holes must be proportional to the relation

$K_A = A_0 / A_1$, where

A_0—the total area of the holes,

A_1—the total area of the walls and the ceiling.

So the coefficient η''_{12} can be found from the equation

$$\eta_{12}'' = (\eta_1 + \eta_{12}' + \eta_{12}'')K_A. \qquad (6)$$

When $A_0 \ll A_1$, Equation (6) can be transformed into the equation

$$\eta' \approx (\eta_1 + \eta_{12}')K_A. \qquad (7)$$

Substituting Equation (7) in Equation (5), we have

$$W_1^{in}/W_2^d = (1+K_A)/(\eta_{12}'/(\eta_1+\eta_{12}')+K_A). \qquad (8)$$

The attenuation of machine noise by the walls and the ceiling can be found as

$$\Delta L_W = 10(\log W_1^{in} - \log W_2^d) = 10\{\log(1+K_A) - \log[\eta_{12}'/(\eta_1+\eta_{12}')+K_A]\} \qquad (9)$$

where

η_1—loss factor of the absorption of the noise by the walls, the ceiling and the floor, it can be accepted equal to absorption coefficients of materials [3].

According to [1]

$$\eta_{12}' = \frac{\tau c A_1}{8\pi f V_1}, \qquad (10)$$

where c—sound velocity in air = $10^{-TL/10}$—Transmission loss factor TL—transmission loss V1—volume of the internal acoustical space, m³

f—sound frequency, Hz.

The transmission loss depends on the sound frequency and the critical frequency f_c of the walls [2],

$$f_c = \frac{c^2}{1.8c_L t},$$
(11)

Where c_L—velocity of longitudinal waves in the barrier (walls, ceiling), m/s t—barrier width, m.

For the walls and ceiling of bricks with t = 0.15 m and f_c = 129 Hz, c_L = 3400 m/s, c = 344 m/s.

When $f < f_c$ the mass law actuates and

$$TL_M = 10\log\left[1+\left(\omega\rho_s/2\rho c\right)\right]-5, \text{ dB}$$

when $f = f_c$

$$TL = TL_M - 37, \text{ dB},$$

when

$$f > f_c$$

$$TL = TL_M + 10\log\eta_W + 10\log\left(\frac{f}{f_c}-1\right)+3, \text{ dB},$$
(12)

where —air density, kg/m³
$_s$—unit area density, kg/m²
$_W$—walls dissipation coefficient (for bricks $_W$ = 0.01).

The total area of the walls and ceiling of the plant is A = 110.5 m², the volume V_1 = 110.05 m³. The area of the holes was estimated as A = 3 m².

The attenuation of sound is calculated using Equations (10) - (14). The results of these calculations are presented in Figure 4. The curve 1 corresponds to the initial conditions. It gives only 15 dB except the zone of critical frequency where the attenuation falls till 5 dB. This almost coincides with the results of noise measurements (Figure 1). The reduction of the area of holes up to 1 m increases the attenuation by 5 dB except in the critical frequency (curve 2). For further increase of the attenuation, two silencers for the suction and exhaust holes were designed and fabricated. The silencers are of absorbing type and were designed using fiberglass and perforated steel sheets.

The attenuation of sound in silencers can be estimated [4] [5] using this equation:

$$\Delta L' = \left[12.6 l \alpha^{1.4} \left(\frac{P}{S} \right) \right] \frac{1}{12}, \text{dB} \tag{13}$$

where l—longitude of silencer, m

—absorption coefficient P—interior perimeter of the duct, m S—interior transverse area of the duct, m².

The absolute value of this sound absorption is

$$q = 10^{0.1\Delta L'}. \tag{14}$$

Taking into account the silencers attenuation, Equation (9) will transform in the following

$$\Delta L_W = 10 \left\{ \log\left(1 + \frac{K_A}{q}\right) - \log\left[\frac{\eta'_{12}}{\eta_1 + \eta'_{12}} + \frac{K_A}{q}\right] \right\}. \tag{15}$$

Using Equation (17) curve 3 and 4 in Figure 4 is calculated. As we see from Figure 4 the introduction of silencers reduces the noise very much except in the critical frequency. For reducing the noise in the critical frequency and overall, the walls and the ceiling are covered with fiberglass of 2 inch thick (curve 4 in Figure 4).

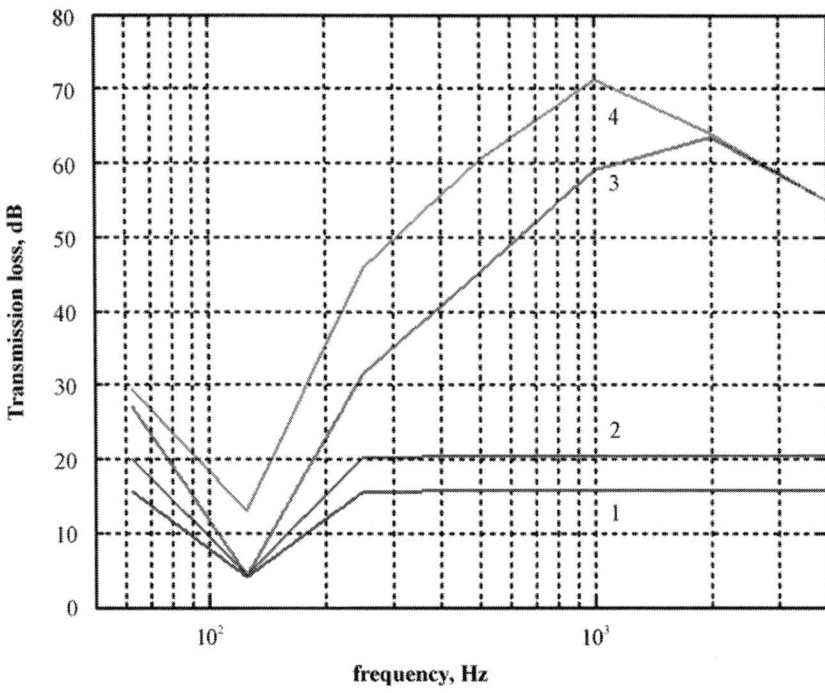

Figure 4: Sound transmission loss for different conditions: 1—initial conditions with the holes of 3 m², 2—the holes area reduced till 1 m², 3—the silencers are installed, 4—the silencers are installed and the walls and ceiling are covered with fiberglass of 2 inch thick.

In Figure 1 the effect of this vibroacoustical design in the reduction of noise level of the plant is presented. It can be seen that the noise level is reduced by 23 dBA at the distance of 15 m away from the source. Now the noise of the plant in the working area is so low that it is practically not heard.

CONCLUSIONS

- The noise measurements of the plant in the university campus show that the noise level exceeds the permissible levels in the offices and classrooms.
- An analysis of the acoustical system of the plant using the SEA method shows that the walls and the ceiling don't have enough

transmission loss because of vast holes at the suction of air and at the exhaust.
- The reduction of the hole areas from 3 m² to 1 m² does not give a sufficient reduction of the noise.
- By using fiberglass coatings for the walls and ceiling and especially designed silencers, a sufficient reduction of noise level (23 dBA at the distance of 15 m from the source) is achieved.

REFERENCES

1. Lion, R. (1975) Statistical Energy Analysis. Theory and Applications. The MIT Press, Cambridge and London.
2. Leo, L. and Beranek, I.L.V. (1992) Noise and Vibration Control Engineering. Principles and Application. Wiley & Sons, Inc., New York.
3. Cyril, M.H. (1995) Manual de Medidas Acústicas y Control del Ruido. Tercera Edición. McGraw-Hill, New York.
4. Faulkner, L.L. (1975) Handbook of Industrial Noise Control. Industrial Press Inc., South Norwalk.
5. Lewis, H.B. and Douglas, H.B. (1994) Industrial Noise Control. Fundamentals and Applications. 2nd Edition. Marcel Dekker, Inc., New York.

Chapter 5

Thermo-Dynamical Analysis on Electricity-Generation Subsystem of CAES Power Plant

Wenyi Liu, Gang Xu, and Yongping Yang

Key Lab of Education Ministry for Power Plant Equipments Conditions Monitoring and Fault Diagnosis, North China Electric Power University, Beijing, China

ABSTRACT

Besides pumped hydropower, Compressed Air Energy Storage (CAES) is the other solution for large energy storage capacity. It can balance fluctuations in supply and demand of electricity. CAES is essential part of smart power grids. Linked with the flow structure and dynamic characteristic of electricity generation subsystem and its components, a simulation model is proposed. Thermo-dynamical performance on off-design conditions have been analyzed with constant air mass flux and constant gas combustion temperature. Some simulation diagrams of curve are plotted too. The contrast of varied operation mode thermal performance is made between CAES power plant and simple gas turbine power plant.

INTRODUCTION

With the development of renewable energy, the grid load regulation problem is becoming more and more serious. Its performance is dependent on energy storage system because of the load fluctuation. Some problems can be solved by electrical energy storage system, for example the peak & off-peak demand of power generation, the improvement of reliability and steadiness of power supply [1] [2].

In CAES technology, air is compressed with a motor/generator using low cost, off-peak electricity and stored underground in caverns or porous media. This is called energy storage subsystem. This pressurized air is released from the ground and then be mixed and burned with gas in a combustor. The hot expanding gases drive a turbo expander and run a motor/generator which, in turn, produces electricity during peak demand periods. This is the other subsystem, electricity generation subsystem. The electricity generation subsystem of CAES includes [3]: isobaric heating process in burner, adiabatic expanding process in turbo expander and isobaric heat release process Main influence factors of practical thermo-dynamic processes including burning efficiency of burner η_b, expansion efficiency of turbo expander and some performance parameters of flowing loss processes. Combined with the characteristics of this subsystem and its components, an off-design condition simulation model is proposed based on unit's modeling system. Equipment's selection and their off-design condition characteristics are analyzed.

SIMULATION MODELING [4] [5]

Turbo expander is key equipment of electricity generation subsystem for CAES. For the sake of utilizing high pressure air released from air storage dome, the air expander can be added before gas expander.

At the same time, the re-heater and recuperator can be used in this subsystem. Figure 1 is the conceptual diagram of electricity generation subsystem. M/G is motor/generator, Air EXP is air expander, Gas EXP is gas expander, REC is recuperator and AS is air storage dome.

The electricity generation subsystem can be divided to some modeling parts, as Figure 2.

Model based on assumptions as follows: 1) air and gas is ideal air, 2) the specific heat of air is constant, 3) the flowing of air and heat transfer between air and wall is steady process, 4) η_b is constant

The equations of these typical parts can be formulated. They include 6 differential equations. With consideration of practical conditions, the heat content of air and gas can be ignored; the compressibility of air and gas in recuperator can be ignored too.

The equation for metal heat storage is Equation (1):

$$\rho_w \cdot \delta \cdot c_p \frac{dT_m}{d\tau} = \alpha_g (\overline{T_g} - \overline{T_m}) - \alpha_a (\overline{T_m} - \overline{T_a}) \tag{1}$$

δ—the thickness of recuperator wall

α_g, α_a—hot end and cold end convection heat transfer coefficients

$\overline{T_a}, \overline{T_g}$ —average temperature of air part and gas part

ρ_w, A_w—density of wall and heat transfer area

Based on $Q = KA\Delta t$, $t = 75$ K and $K_0 = 100$ W/ (m²·K) on design condition.

The density of recuperator material is 7800 kg/m³, specific heat $c_p = 460$ J/ (kg·K), $\alpha_g = \alpha_a = \alpha$, so the time constant can be derived as Equation (2) on design condition:

$$\frac{\rho_w V_w}{\alpha A_w} c_p = \tau_c = 3588000 \delta / a \tag{2}$$

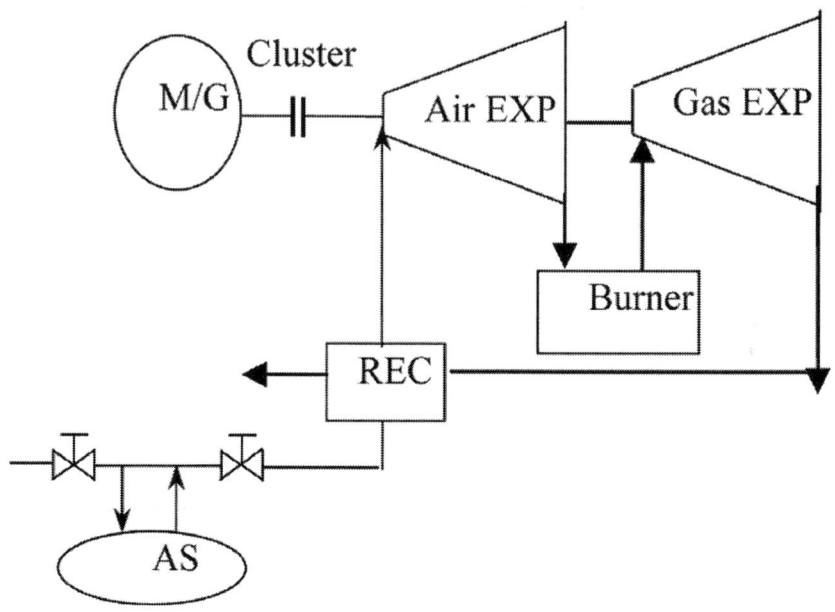

Figure 1: The conceptual diagram of electricity generation subsystem.

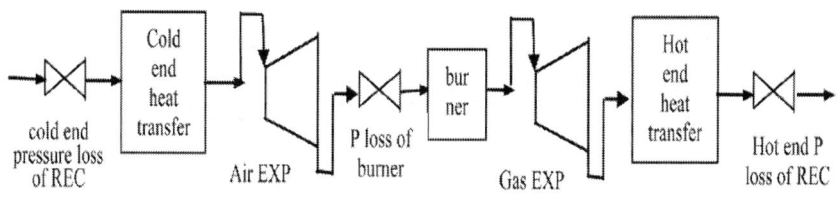

Figure 2: Typical modeling parts of electricity generation subsystem.

The burning efficiency of burner can be expressed as Equation (3).

$$\eta_b = f(\frac{p^{1.75} A_{max} D_{max} e^{T/300}}{G_a})$$

(3)

P, T—inlet pressure and inlet temperature of burner

A_{max}, D_{max}—maximum cross-section area and its diameter

The off-design equations of recuperator is Equation (4)

$$\frac{\sigma}{\sigma_0} = 1/[\sigma_0 + (1-\sigma_0)(\frac{G_a}{G_{a0}})(\frac{K_0}{K})] \quad (4)$$

Heat transfer coefficient K is associated with recuperator type and air flux, so the off-design formula can be expressed as Equation (5)

$$\frac{K_0}{K} = (\frac{G_{a0}}{G_a})^{0.8} (\frac{\overline{T_{a0}}}{\overline{T_a}})^{0.06} (\frac{\overline{T_{g0}}}{\overline{T_g}})^{0.16} \quad (5)$$

OFF-DESIGN CONDITIONS OPERATION SIMULATION [4]

PG9171E (GE) can be taken as simulation object, main parameters as Table 1. Some values are as follows: 1) Relative internal efficiency of gas expander is 0.905 ($k = 1.33$) and air expander is 0.88. 2) The pressure losses in burner, cold end of recuperate and hot end of recuperator are 3%, 1% and 3% respectively. *LHV* of fuel is 41,960 kJ/kg, working consuming for compressing air $P_c = 617.65$ kJ/kg [4], outlet air pressure of air storage dome is 61bar, temperature is 30°C.

The additional formulations are as follows (6)-(9):

$$0 = c_{p1} \cdot G_a \cdot (t_1 - t_2) - c_{p1} \cdot G_a \cdot (t_1 + 273)(1 - \pi_1^{-\frac{k_1-1}{k_1}}) \cdot \eta_a \quad (6)$$

$$0 = \eta_b \cdot LHV \cdot G_f - (G_a + G_f) \cdot c_{p2} \cdot t_3 + c_{p1} \cdot G_a \cdot t_2 \quad (7)$$

$$0 = c_{p2} \cdot (G_a + G_f) \cdot (t_3 - t_4) - c_{p2} \cdot (G_a + G_f) \cdot (t_3 + 273)(1 - \pi_2^{-\frac{k_2-1}{k_2}}) \cdot \eta_t \quad (8)$$

$$0 = \frac{t_1 - 30}{t_4 - 30} - 0.85 \quad (9)$$

t_1, t_2, —inlet T and outlet T of air expander.
0.85—the heat transfers efficiency of recuperator.
- Simulation calculations on design air flux, additional equation are as follows (10)-(12) [6] [7]:

$$\tilde{b} = [c^{b_5} \cdot (c^g + c^l) \cdot (\tilde{t}^3 - \tilde{t}^4) + c^{b_l} \cdot c^g \cdot (\tilde{t}^1 - \tilde{t}^5)] \tilde{w}^{\text{m}} \cdot \tilde{w}^g \quad (10)$$

$$\frac{\eta_a}{\eta_{a0}} = \sqrt{\frac{h_{as0}}{h_{as}}} (2 - \sqrt{\frac{h_{as0}}{h_{as}}}) \quad (11)$$

$$\frac{\eta_t}{\eta_{t0}} = \sqrt{\frac{h_{ts0}}{h_{ts}}} (2 - \sqrt{\frac{h_{ts0}}{h_{ts}}}) \quad (12)$$

The calculation results are shown as Figures 3-5. Some conclusions can be drawn from figures. With the load lowered, the internal efficiency of air expander and gas expander is lowered obviously, heat rate of generating electricity is lowered, electricity rate of generating electricity is improved obviously, the energy transformation coefficient is lowered obviously too.

- Simulation calculations on design inlet temperature of gas expander.
- Some results are seen as Figures 6-8. Some conclusions are: with the load lowered, the internal efficiency of air expander and gas expander is constant approximately, heat rate and electricity rate of generating electricity are improved, and the energy transformation coefficient is lowered a little.
- Some results are seen as Figures 6-8. Some conclusions are: with the load lowered, the internal efficiency of air expander and gas expander is constant approximately, heat rate and electricity rate of generating electricity are improved, and the energy transformation coefficient is lowered a little.
- Contrast with off-design gas turbine performance.
- On the basis of off-design performance of gas turbine, using former selected gas turbine, the performance contrast diagram of gas turbine and electricity generation subsystem is shown in Figure 9 with the load scope 100% - 40%.

Table 1: Main performance parameters of gas turbine

Type	Pe (kW)	e (%)	Pressure ratio	Ga (kg/s)	Initial T/t3 (°C)	Leakage T/t4 (°C)	Leakage air flus G4 (kg/s)
PG9171E	123400	33.8	12.3	403.7	1124	538	412.4

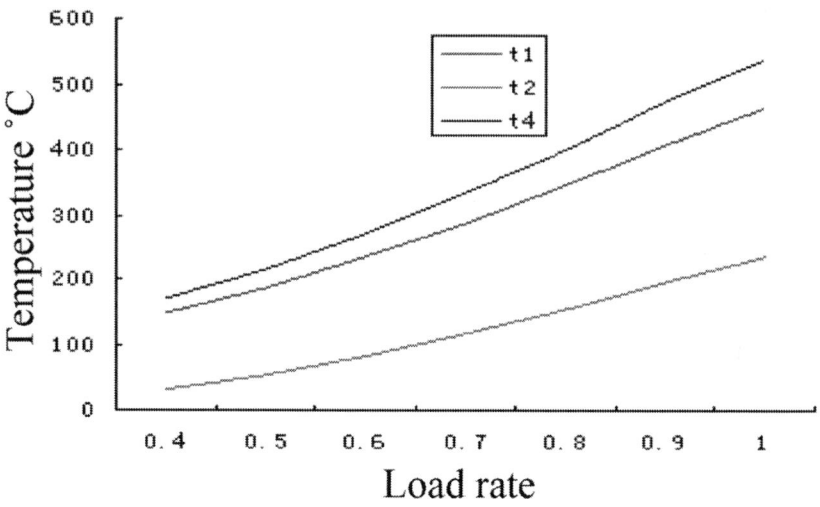

Figure 3: t_1, t_2, t_4 varying with load rate on design air flux.

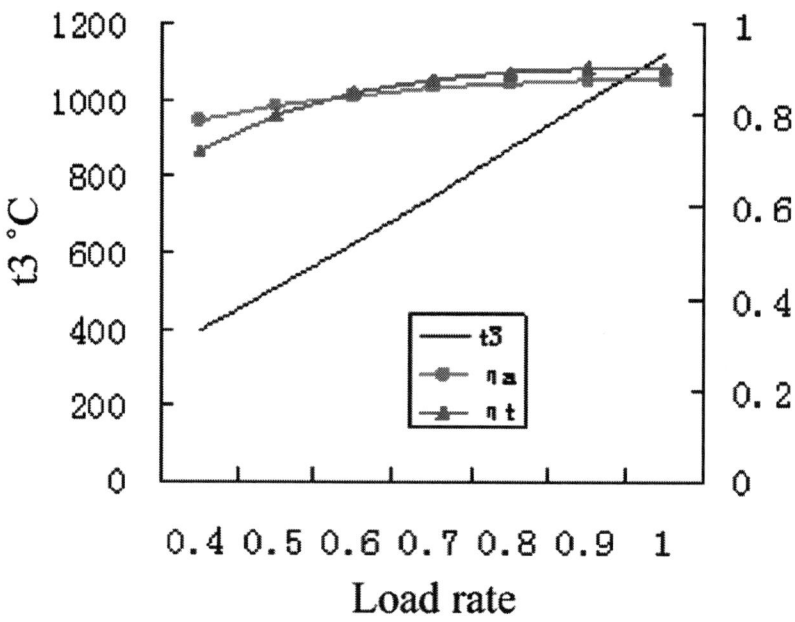

Figure 4: t_3, η_a, η_t varying with load rate on design air flux.

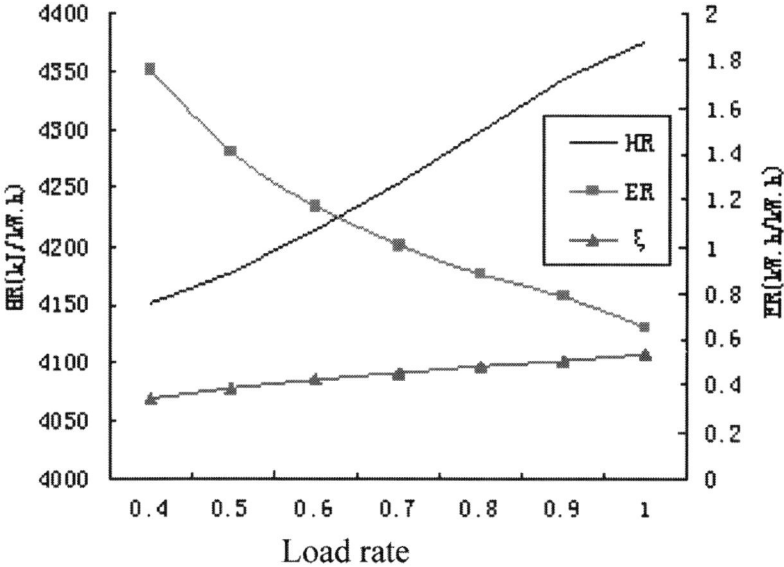

Figure 5: HR, ER, ξ varying with load rate on design air flux.

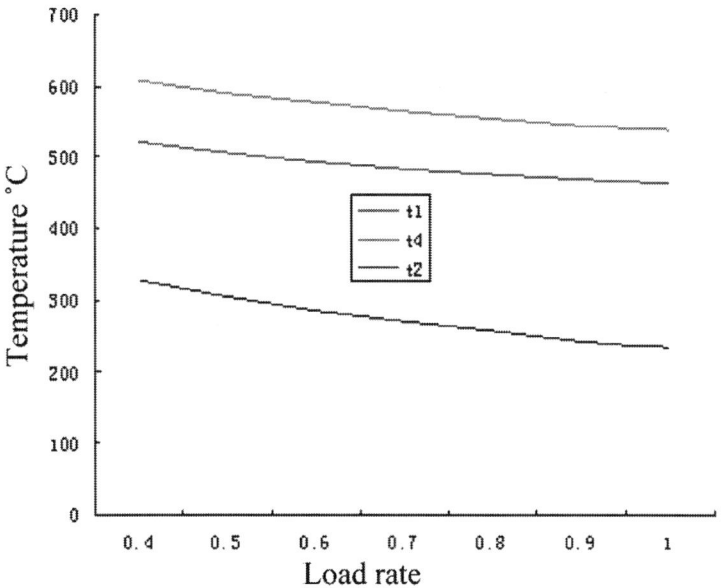

Figure 6: t_1, t_2, t_4 varying design inlet temperature.

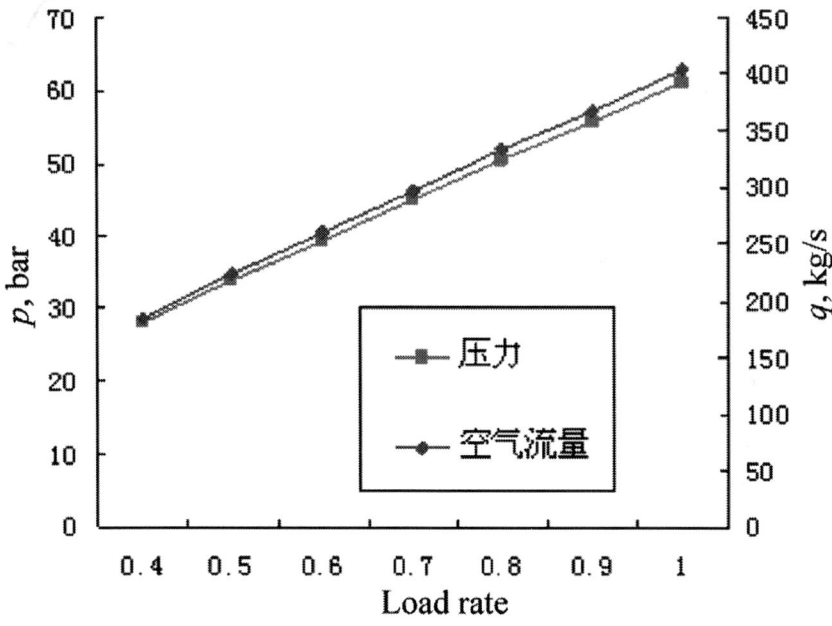

Figure 7: t_3, η_a, η_t varying with load rate on design inlet temperature.

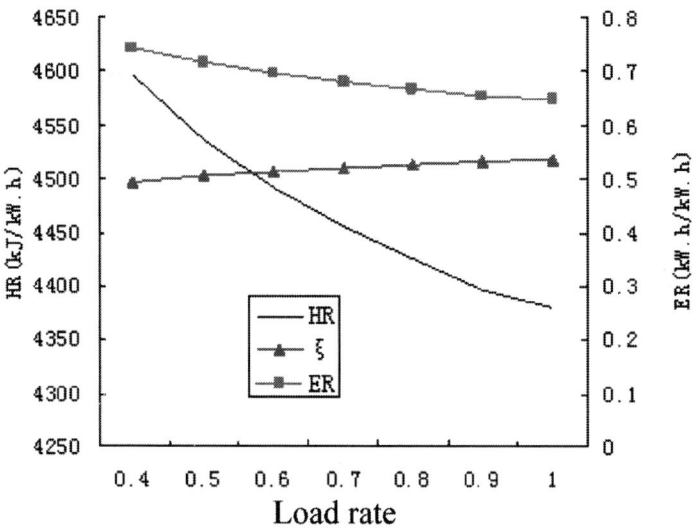

Figure 8: HR, ER, ξ varying with load rate on design inlet temperature.

It is obvious that thermal efficiency is varying from 33.8% to 0% of gas turbine power plant when load rate varying among 100% - 40%. But energy transform coefficient varies from 53.6% to 34.3% on design air flux, from 53.6% to 49.5% on design inlet temperature.

CONCLUSIONS

- The electricity generation subsystem of CAES is divided into three processes; the difference between practical process and ideal process is described and analyzed in this paper.
- Combined with the characteristic of electricity generation subsystem and its components, a subsystem simulation model is proposed based on unit's modeling system.
- The simulation diagrams of off-design condition are plotted. The conclusion is that thermal efficiency is varying from 33.8% to 0% of gas turbine power plant when load rate varying among 100% - 40%. But energy transform coefficient varies from 53.6% to 34.3% on design air flux, from 53.6% to 49.5% on design inlet temperature.

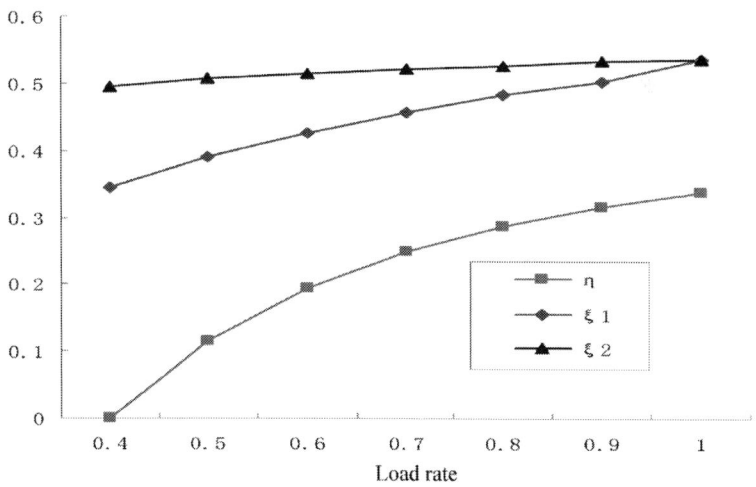

Figure 9: η, ξ_1, ξ_2 varying with load rate. H—thermal efficiency of gas turbine; ξ_1—energy transform coefficient on design air flux; ξ_2—energy transform coefficient on design inlet temperature.

ACKNOWLEDGEMENTS

The paper is supported by National Nature Science Fund of China (No. 51276059).

REFERENCES

1. Cavallo, A.J. (2011) Energy Storage Technologies for Utility Scale Intermittent Renewable Energy Systems. *Transactions of the ASME*, 123, 387-389.
2. Baker, J.N. and Collison, A. (1999) Electrical Energy Storage at the Turn of Millennium. *Power Engineering Journal*, 107-112. http://dx.doi.org/10.1049/pe:19990301
3. Liu, W.-Y., Yang, Y.-P. And Song, Z.-P (2005) Optimization and Performance Simulation of Different CAES Systems. *Journal of Engineering Thermophysics*, 26, 25-28.
4. Liu, W.-Y. Simulation Analysis of Thermal Performance for Compressed Air Energy Storage (CAES) Power Plant Doctor Dissertation of North China Electric Power University, Beijing, 19-36.
5. Ni, W.-D., Xu, X.-D., Li, Z., *et al.* (1996) Some Problems of Thermal Power System Model and Control. 1st Edition, Beijing Science Press.
6. Lin, R.-M., Liu, R.-T., Jin, H.-G., *et al.* (2004) Gas Turbine Selection Problems of Combined Cycle Power Plant. *Gas Turbine Technology*, 9, 6-13.
7. Zhang, N. and Cai, R.X. (2000) Off-Design Condition Typical Analytical Characters of Single-Axis Gas Turbine Combined Cycle. *Journal of Engineering Thermophysics*, 21, 529-532.

Chapter 6

The Optimal Steam Pressure of Thermal Power Plant in a Given Load

Yong Hu, Ji-zhen Liu, De-liang Zeng, Wei Wang, and Ya-zhe Li

North China Electric Power University, State Key Laboratory of Alternate Electrical Power System with Renewable Energy Sources, Beijing, China

ABSTRACT

As the large change of the grid load, many large capacity units of our country had to change the load in order to meet the gird need. When a thermal power plant receives a given load instruction from the grid, it is necessary to set an optimal steam pressure to maintain the high efficiency of the plant. In the past optimization methods, during the process of calculation, the output of the turbine often changed, it was hard to maintain the output constant. Therefore, in combination with the theory of variable condition of turbine, calculation of governing stage and the matrix equation of thermal power system, an optimization method were put forward and an optimal solution was got in a given load.

INTRODUCTION

As the development of economy in china, consumption level of the people have enhanced, which leads to a large proportion of electricity power is consumed in our daily life, resulting in the difference between peak and valley of grid load increased year by year. And in our country, large-capacity thermal power plants have a large percentage in the total installed capacity of power plants, which makes the large-capacity units with a basic load have to participate in the load regulation. The units have to deviate from the original design condition and even run in the low load area for a long time, which makes thermal efficiency of the units decrease greatly. Among the factors that affect the thermal efficiency of power plant, only the running modes and operating parameters can be adjusted by operating personnel. Therefore the research of thermal power plants in the off-design condition is of great significance in the selecting of running modes and operating parameters.

In a given load, when the unit runs in a high steam pressure, the ideal enthalpy drop of turbine will increase and the outlet pressure of feed-water pump will rise simultaneously. In order to maintain the given unit load, it is necessary to reduce the steam flow rate through de-creasing the opening degree of regulating valves; this will increase the throttling loss of the governing stage. When the unit runs in a low steam pressure, the theoretical thermal efficiency of the unit will reduce, but the lower steam pressure will make the governing stage to maintain higher internal efficiency, and the outlet press-sure of feed-water pump will decrease. In order to main-taint the unit load, it has to enlarge the opening degree of regulating valves to increase the steam flow rate. There-fore, in tracking of the grid given load, the unit usually deviates from the designed condition, how to select the optimal steam pressure and the running mode has a great influence on the interest of the power plant.

In the traditional method of the pressure optimization, it usually assumed the steam pressure was approximately proportional to the steam flow, when the steam pressure changed, it calculated the steam flow firstly and then calculated the back pressure of governing stage according to the Flu gel formula [1], carried on variable condition calculation of the governing stage and the whole turbine. Finally it determined the efficiency of the unit under the changed steam pressure [2]. But in the practical operation of the thermal power plant, it must guarantee the load of the unit equal to the instruction from the grid, when the steam pressure gets higher, it needs to decrease the opening degree of the regulating valves, lower the steam flow to ensure the stability of the load, and vice versa. In the traditional method, due to the approximate proportional relationship between steam pressure and the flow, it leads to the load changed in proportion, not in-variable. In some other literatures, in order to ensure the load unchanged, it iteratively calculated the steam flow using the turbine power equation [3], which ignored the characteristic of the governing stage and caused the deviations of the results

Therefore, on the base of variable condition calculation method of governing stage and variable condition theory of turbine, using thermal economic matrix equation [4], in order to solve the problems mentioned above, a new calculation method of optimal steam pressure in a given load was put forward, the optimal steam pressure and running modes was got under different loads.

MODEL OF STEAM PRESSURE OPTIMIZATION

Calculation of the Governing Stage

In the variable condition calculation of the governing stage, the steam flow through the fully opened regulating valves and the partly opened valve can be expressed as:

$$G'_n = 0.648 \cdot \frac{A'_n}{10 \cdot \sqrt{p_0 \cdot v_0}} \cdot \mu' \cdot p_2 \tag{1}$$

$$G''_n = 0.648 \cdot \frac{A''_n}{10 \cdot \sqrt{p_0 \cdot v_0}} \cdot \mu'' \cdot p_2 \tag{2}$$

Then main stream flow rate can be expressed as:

$$G_n = G'_n + G''_n \tag{3}$$

In which G_n is the steam flow through the fully opened valves; G'_n is the steam flow through the partly opened valve; A'_n is the flow area of fully opened valves; A''_n is the flow area of partly opened valve; P_0 is the pressure of main stream; V_0 is the specific volume of main steam; μ' μ'' is the function of P_2/P_2, P_2/P_0; P_0 is the steam pressure behind fully opened valve; P_0 is the steam pressure behind partly opened valve; P_2 is the back pressure of governing stage; η the efficiency of governing stage; x_a is the speed ratio of governing stage.

In general, when the steam flows through the fully opened valves, the throttle loss is smaller, so it can be assumed $P_0 = 0.95\, p_0$; when the steam flows through the partly opened valve, the opening degree of partly opened valve is x ($x \in [0,1]$), since the annular chamber after the nozzle is in communication with each other, the steam pressure P_1 behind the nozzle of each nozzle group are the same, the steam pressure behind the un-opened regulating valve (i.e., the pressure before the nozzle of unopened valve) is also equal to P_1. When the opening degree of valve x gradually changes from 0 to 1, the pressure P''_0 behind partly opened valve will change from P_1 to $0.95\, p_0$ In order to facilitate the calculation $P_0 = (0.95\, p_0 - p_1)\, x + p_1$ is assumed (this assumption is only convenient to calculate P''_0, it has no effect on the optimization results). So when the opening degree of all the regulating valves is known, the main steam flow can be expressed as

$$G_n = f(p_0, x, p_2) \tag{4}$$

Therefore, the main steam flow can be determined by P_0, x and P_2, then on basis of the variable condition calculation of governing stage, the steam enthalpy of governing stage h_{tj} can be got.

Calculation of the Intermediate Stage and Last Stage

In the variable condition calculation of turbine, because the flow area of intermediate stage is constant, when the load of the unit is changed, if the variation of temperature before all stages is ignored, the pressure before intermediate stage is proportional to the steam flow of this stage, so pressure ratio is invariant, the efficiency of intermediate stage is unchanged, the ideal enthalpy drop of each stage is also unchanged [5]. Therefore, when the parameters of governing stage are known, the steam enthalpy of each extraction point can be expressed as:

$$h_{(i+1)1} = h_{i1} - (h_{i0} - h_{(i+1)0}) \tag{5}$$

In which h_i is the steam enthalpy of ith stage; subscript 0 represents the designed condition; subscript 1 represents the variable condition.

For the last stage of steam turbine, we calculated from the last stage to the prior stage, found the superheated steam extraction point and set it as ith stage. The steam after the ith stage does adiabatic expansion in the turbine, so the entropy is constant. Combined with the steam pressure of extraction point, the ideal steam en-thalpy of this stage could be got, according to (6), we could get the steam enthalpy of this stage and calculated one stage by one stage until to the last stage.

$$h_{(i+1)1} = h_{i1} - \eta_{i,i+1}(h_{i1} - \tilde{h}_{(i+1)1}) \tag{6}$$

In which $\tilde{h}_{(i+1)1}$ is the ideal steam enthalpy of (i+1) stage, $\eta_{i,i+1}$ is the efficiency of stage (i+1)

Calculation of the Boiler Feed-Water Pump Turbine

When the main steam pressure and flow rate change, the output of Boiler Feed-Water Pump Turbine (BFPT) will change too. Therefore, the influence of BFPT on the thermal efficiency cannot be ignored.

From the outlet of feed-water pump to the main steam valve, the phase of working fluid changed. In this process, there exists the loss of resistance along the way and the loss of local resistance [5], both loss can be expressed as:

$$\Delta p = \zeta \cdot \rho \cdot \frac{C^2}{2} \tag{7}$$

In which Δp is the pressure drop; ρ is the average density of fluid; C is the flow rate of fluid; ζ is the loss coefficient which depends on the characteristic of pipe. We use subscript d represent the parameters of design-condition, then the outlet pressure of feed-water pump can be expressed as:

$$P_{p2} = P_0 + (\frac{G_n}{G_{nd}})^2 \cdot \frac{\rho_d}{\rho} \cdot (P_{p2d} - P_{0d}) \tag{8}$$

In which p_{p2} is the outlet pressure of feed-water pump; P_0 is the main steam pressure; G_n is the main steam flow. When the feed-water flows through the pump, the pressure of feed-water will rise because of the working of pump, this will make the feed-water enthalpy rise. This process can be regarded as isentropic flow [6], so the enthalpy- rise of feed-water can be expressed as:

$$\Delta h = \frac{\overline{v} \cdot (P_{p2} - P_{p1})}{\eta_p} \tag{9}$$

In which p_{p1} is the inlet pressure of feed-water pump; $\overline{v}$ is the average specific volume of feed-water; η_p is the efficiency of feed-water pump. According to the law of conservation of energy, the extraction flow for BFPT can be got.

$$D_{BFPT} = \frac{G_n \cdot (p_{p2} - p_{p1}) \cdot \overline{v}}{(h_4 - h_{pc}) \cdot \eta_p \cdot \eta_j} \qquad (10)$$

In which h_4 is the inlet steam enthalpy of BFPT; h_{pc} is the exhaust enthalpy of BFPT; η_j is the efficiency of BFPT.

OPTIMIZATION METHOD OF STEAM PRESSURE IN A GIVEN LOAD

Optimization Method

In order to maintain the output of the unit and overcome the shortcomings of traditional optimization methods, in the process of optimization, we adopted the sequential calculation method, combining with assumption, verification and iterative adjustment. If the load and an initial steam pressure were given, we could get the steam flow, opening degrees of regulating valves, back pressure of governing stage and thermal efficiency of the unit, and then a unique mapping relationship was formed among them.

Step1. According to the load instruction, use Figure 1 to determine the feasible range of steam pressure [7] and the number of fully opened valves.

Step2. Assume a certain back pressure of governing stage and the degree of partly opened valve to determine the main steam flow, the enthalpy and temperature of governing stage.

Step3. According to the enthalpy and temperature of the governing stage, carry on the variable condition calculation of intermediate stage and last stage, get the out-put of turbine.

Step4. Judge the parameters of governing stage using (11). Equation (11) is the Flugel formula [1]. If the equation does not hold, adjust the back pressure of governing stage, and then go to step 2.

$$\frac{G_n}{G_{nd}} = \sqrt{\frac{p_2^2 - p_c^2}{p_{2d}^2 - p_{cd}^2}} \cdot \sqrt{\frac{T_{2d}}{T_2}}$$

(11)

Step5. Judge the output of turbine. If the output of turbine is not equal to the load instruction, adjust the degree of partly opened valve and go to step 2.

The flow chart of the pressure calculation is shown in Figure 2.

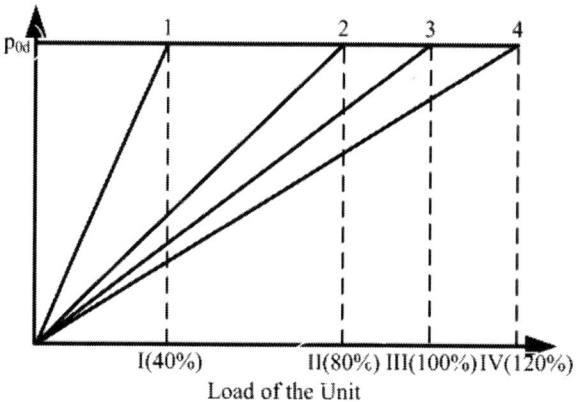

Figure 1: The feasible range of steam pressure.

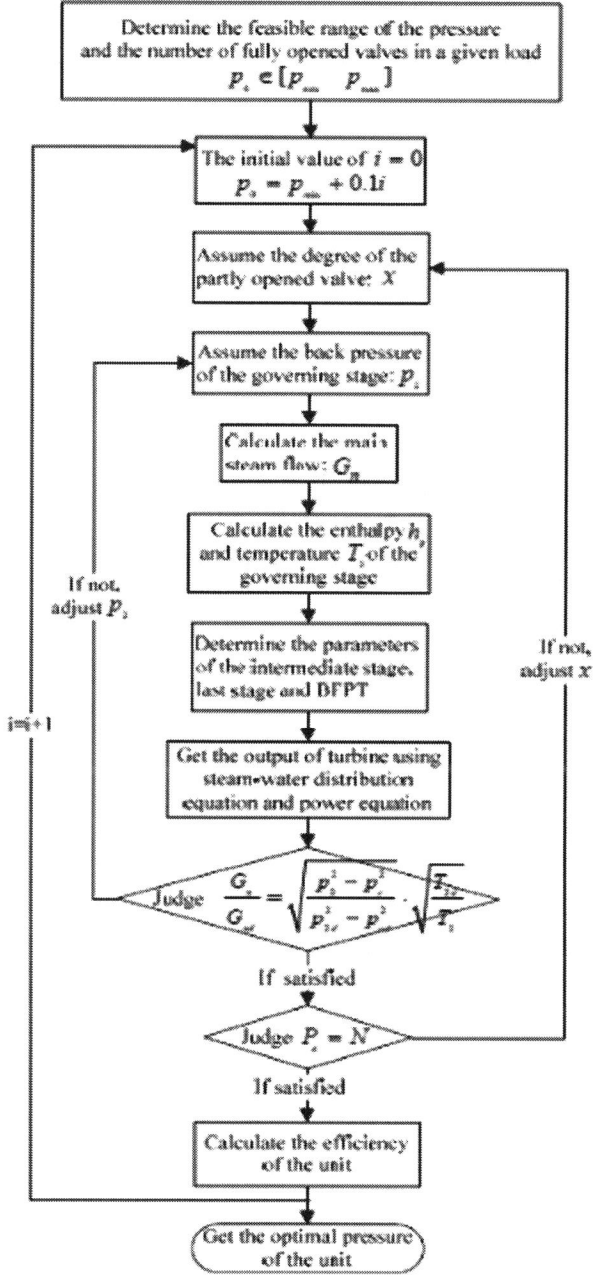

Figure 2: The flow chart of the optimization method

Application Examples

We took the Oriental steam turbine N1000-25.0/600/600 as an example, the impact of the overlap of regulating valves was not considered and we ignored the influence of environmental factors on the thermal economy of the unit. In the ideal condition of 100% load, there were three regulating valves fully opened and one valve closed.

First, we took the 100% THA condition as an example, analyze and validate the optimization method, the results were shown in Table 1. In the 100% THA condition, in order to maintain the unit load, as the decline of main steam pressure, the regulating valves had to be opened larger to increase the main steam flow, and the power consumed by feed-water pump was decrease too. The thermal efficiency of the unit was decline as the steam pressure became lower. But when the main steam pressure reduced to 22.76 Mpa, four regulating valves were all fully opened, the throttling losses was least at this moment, so the efficiency of the unit rebounded a little. From the dates of Table 1, the variation tendency of pressure and efficiency consistent with the theoretical analysis, so this method can be used to optimize other conditions of the unit.

Figure 3 shows the thermal efficiency change as the number of fully opened valves change from 2 to 4 in different load. As the increasing of the opening degree, the main steam pressure dropped, the thermal efficiency declined, but in the fully opened points, there existed a local optimal point.

Table 1: The analysis of efficiency in a design condition

Steam Pressure (Mpa)	Main Steam Flow(t/h)	Degree of Regulating Valves	Thermal Efficiency	The Energy Consumption of Feed-water Pump (MW)
25.00	2850.95	75.00%	49.0875%	34.6134
24.00	2880.56	79.25%	48.8659%	33.7388
23.60	2885.50	83.75%	48.8319%	33.2834
23.20	2889.22	90.50%	48.8106%	32.8091
23.00	2889.43	94.50%	48.8095%	32.5484
22.76	2888.78	100.0%	48.8159%	32.2231

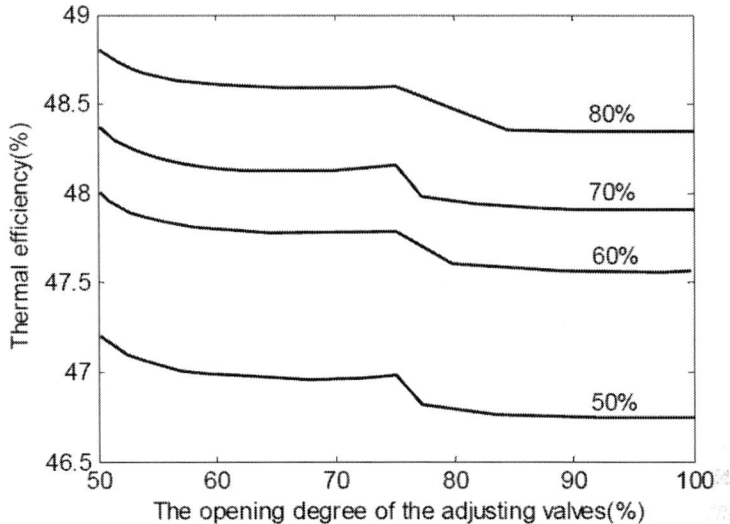

Figure 3: The relation between thermal efficiency and valve opening.

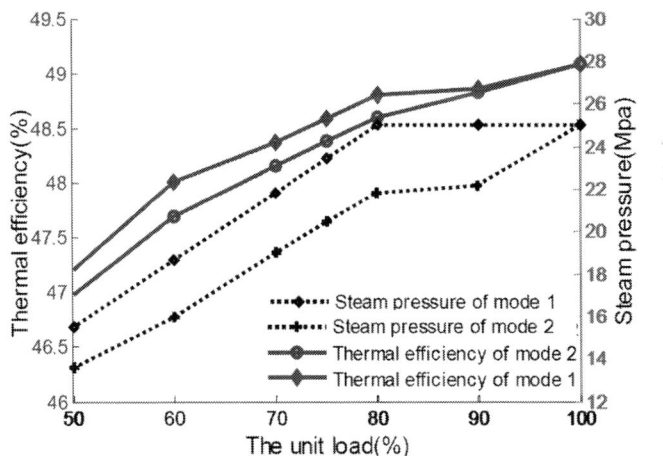

Figure 4: The comparison curves of the two modes.

Figure 4 shows the efficiency of the unit in different sliding pressure operation mode. In mode 1, the unit took a fixed pressure operation mode with 25Mpa steam pressure first, when the output of the load decreased to the 80%, the unit took the sliding pressure operation mode with 2 regulating valves fully opened. In mode 2, the unit took the sliding pressure operation mode with 3 regulating valves fully opened beginning from the 100% THA condition. It can be seen from the picture, the efficiency of mode 1 was higher than mode 2, especially in the low load region.

CONCLUSIONS

Basing on the variable condition calculation method of governing stage and variable conditions theory of turbine, a new optimization method was put forward for the optimal operation of thermal power plant, and we took the Oriental steam turbine as an example, got the optimal steam pressure of different load and the optimal sliding pressure curve. Therefore, operating personnel can adopt this method, combined with the characteristic of the unit and the factor of environment, drawing the optimal pressure curves, which can be used as a reference in the practical operation.

ACKNOWLEDGEMENTS

This work was supported by the National Basic Research Program of China ("973" Project) (Grant No. 2012CB- 215203) and the National Natural Science Major Fund Project (Grant No. 51036002)

REFERENCES

1. C. F. Zhang and Y. H. Cui, "The Distinguishing Theory of Critical State of Turbine and Improved Flugel Formula," *Science in china Series E*, Vol. 33, No. 3, 2003, pp.264-272.
2. L. X. Zhou and M. Hua, "Method for Calculating Main Steam Pressure and Heat Rate Correction Curves Under Off-design Operating Conditions," *Journal of Engineer-ing for Thermal Energy and Power*, Vol. 26, No. 3, 2011, pp. 351-353.
3. Z. P. Yang and Y. P. Yang, "Sensitivity Analysis on Energy Consumption of Exhaust Steam Pressure of 1000 MW Steam

Turbine Unit, " *East China Electric Power*, Vol. 39, No. 2, 2011, pp. 2064-2067

4. S. L. Yan and C. F. Zhang, "The Steam-Water Distribu-tion General Matrix Equation of Thermal System for the Coal-Fired Power Unit," *Proceedings of the CSEE*, Vol. 20, No. 8, 2000, pp.69-73.
5. N. Zhao, "The Research of the Relation between Pressure Rate and Flow and the Target Value of the Thermal Pa-rameters under Variable Working Conditions of Steam Turbine," *North China Electric Power University*, 2008.
6. P. Li and M. Hua, "Research on Target Value of BFPT Parameters for Sliding Pressure Operation Unit," *Thermal Turbine*, Vol. 39, No. 4, 2010, pp. 251-254.
7. C. F. Zhang and H. J. Wang, "Quantitative Research of Optimal Initial Operation Pressure for the Coal-fired Power Unit Plant," *Proceedings of the CSEE*, Vol. 26, No. 4, 2006, pp. 36-40.

Chapter 7

Practical Implementation of Safety Verification in LNG Production Facilities

Achint Rastogi and Hossam A. Gabbar

`Faculty of Energy Systems and Nuclear Science, University of Ontario Institute of Technology, Oshawa, Canada

ABSTRACT

Many energy and production facilities are operating without clear formal safety requirements, which are considered the base for good process safety management practices. Safety requirements are typically specified during process design based on identified hazard scenarios. This paper proposes a practical framework and methods to systematically synthesize safety requirements based on qualitative and quantitative fault and hazard scenarios. Our aim will be to design a proper safety verification framework which would provide some guidelines regarding the sequence of steps to be taken in the plant for the verification of the safety of that plant. The objective of this paper is to show how the safety verification techniques meet the safety requirements of any production plant. We will clarify Safety Life Cycle and the detailed steps for safety design and verification and also analyze current practices and challenges of safety verification in instrumented/non-in- strumented systems. We will also develop possible activity model for safety verification process and will propose safety requirements representation that will facilitate safety verification. Case study of experimental setup is used to demonstrate the proposed framework, which will support safety design and verification.

INTRODUCTION

The ultimate goal of any organization is to execute all activities so as to achieve a desired level of safety as efficiently and effectively as possible. Governmental safety regulations and international standards all support this goal, with varying degrees of clarity [1]. As we all know, Safety is an important task in chemical plants and plays a significant role throughout the whole design process [2]. Safety is of paramount importance in any industrial plant, be it an LNG plant, production plant or any other production related facility. Lack of safety may lead to hazardous events severely affecting human life, plant and animal life and environmental balance. This paper presents an integrated framework for safety control design based on independent protection layers and defence in-depth concepts. Safety control systems are designed and evaluated in view of safety requirement specifications and corresponding safety rules and constraints are mapped to protection layers or barriers. The proposed safety control design framework can be applied on energy and nuclear power plants, smart grids, oil & gas production plants, or other manufacturing plants. Thus for production facilities, it is necessary to provide a safe atmosphere by proper implementation of safety verification techniques, proper safety instrumented systems and frameworks for safety design of energy and production plants. Verification is the evaluation of an implementation to determine that applicable safety-critical requirements for any plant and its operations are met. The verification process ensures that the design solution meets or exceeds all validated safety requirements.

A verified system shows measurable evidence that it complies with the overall system safety needs by incorporating an efficient safety verification framework.

LITERATURE REVIEW

Accidents happened in the past and are still happening today. If proper measures are not taken, they will continue to happen in the future too. Going through some of the literature, we can easily find that the root cause of all the accidents is lack of a proper safety framework. There is no proper framework for safety verification. Safety Standards and Verification tools are present, but the proper communication between them is absent. A proper framework which links the initiation of a hazard (*i.e.* a fault), safety measures to be adopted (to prevent the propagation of a fault) and verification is missing in the process industry. Our aim will be to design a proper safety verification framework which would provide some guidelines regarding the sequence of steps to be taken in the plant for the verification of the safety of that plant.

Background

Major industrial accidents, like the ones which occurred in Bhopal (India), Dronka (Egypt), Texas City (USA), Three Mile Island (Pennsylvania, USA), Chernobyl (Ukraine), etc. are vivid reminders of the destruction that can occur due to inadequate safety measures. Huge losses of human life, immense environmental pollution, and large capital costs were involved in those accidents.

Unfortunately, extremely serious accidents still happen today. Though modern safety practices include the application of a large number of safeguarding measures, many accidents (refer Table 1) in the process industries are still happening today. These past accidents and the experiences gained from them have led to the development of many technical solutions, like the use of Safety Instrumented Systems (SIS) and Emergency Shutdown Systems (ESS) [3]. In order to implement these technical solutions, numerous safety related standards, like IEC 61508 [4], IEC61511 [5], ISA96 [6], etc. have been written and compliance with these standards is considered a good engineering practice. Compliance with these standards, however, did not prevent several major accidents. As a result of the continuously growing complexity of both industrial processes and the related safety instrumented systems, it appears that new kinds of problems have arisen [7,8].

Table 1: Ten major onshore accidents, worldwide (on the basis of fatalities)

S. No.	Accident Date	Location	Material Name	No. of Fatalities	No. of Injuries
1	3/12/1984	Bhopal (India)	Methyl Isocyanide	>2000	>170,000
2	2/11/1994	Dronka (Egypt)	Aircraft Fuel	>580	N.A
3	19/11/1984	San Juan Ixhuatepec (Mexico)	LPG	>500	2500
4	23/12/2003	Gao Qiao (China)	Natural Gas, Hydrogen Sulphide	243	4000 9000
5	19/12/1982	Tacoa (Venezuela)	Fuel Oil	>153	500
6	14/9/1997	Visakhapatnam (India)	LPG, Crude Oil, Kerosene, Petroleum Products	56	20
7	24/1/1970	Semarang (Indonesia)	Kerosene	50	N.A
8	6/1/1998	Xingping (China)	Nitrogen	50	100
9	24/3/1992	Dakar (Senegal)	Ammonia	41	403
10	19/1/2004	Skikda (Algeria)	LNG	23	74

Root Cause of Accidents

A study on the causes of these incidents and accidents showed that there are some serious problems regarding the quality of information on accidents and the related technical solutions. Hence, adequate control of the quality of safety related information is of huge importance if we want to achieve an acceptable safety level. Also there is a lack of a clear framework which will ensure that the safety standards are also met in practice. This leads to the development of the proposed safety verification framework.

Since last decades, industrial processes are becoming more and more complex [9]. Expanding product and production requirements led to further optimization of the concerned processes. Due to continuously increasing competition, the necessity for increased productivity force process installations to operate to their limits. At the same time, a growing number of different semi manufactured products put a high demand on the flexibility of the process installations, resulting in several different applications. Dedicated instrumentation, which also makes process control more and more complex, is expected to control and safeguard these processes. As a consequence of the growing complexity of the process installations, the control instrumentation, and safeguarding instrumentation, safety related business processes have become even more difficult to manage [10,11].

Fortunately, during the last decades, the process industry has witnessed much improvement. Thorough investigations of accidents have resulted in specific hazardous event prevention with regard to process installations. Consequently, many new safeguarding measures have been developed and are implemented. However, at the same time it has become extremely difficult to acquire a comprehensive view of the entire processes, instrumentation and installations. Due to this growing complexity and an ever expanding process capacity, the potential for serious accidents have heavily increased.

Process Safety Management (PSM) is term frequently used to cover the set of safety related operational activities and processes, which results in a specific safety performance of a process installation. The British Health and Safety Executive (HSE) performed a comprehensive study and clearly illustrated that inadequate process safety management is the most essential factor that contributes to the number of hazardous events [12]. The extent to which failures contributed to explosions in gas fired plants in 1997 were investigated by the HSE. These failures were categorized into four groups (see Figure 1):

- Equipment related failures, such as a manufacturing failure, design faults, or incorrect specification.
- The lack of equipment and equipment, which should have been fitted to the plant, but was not.
- Poor maintenance and incidents resulting directly from poor maintenance/ commissioning.
- Inadequate process safety management.

Other examples of the causes of major industrial incidents are illustrated by Bradley [13]. He found out that 10% of all the investigated failures are contributed by manufacturing and equipment failures. Operating errors, management errors, design/specification errors, and maintenance errors are the remaining contributing factors.

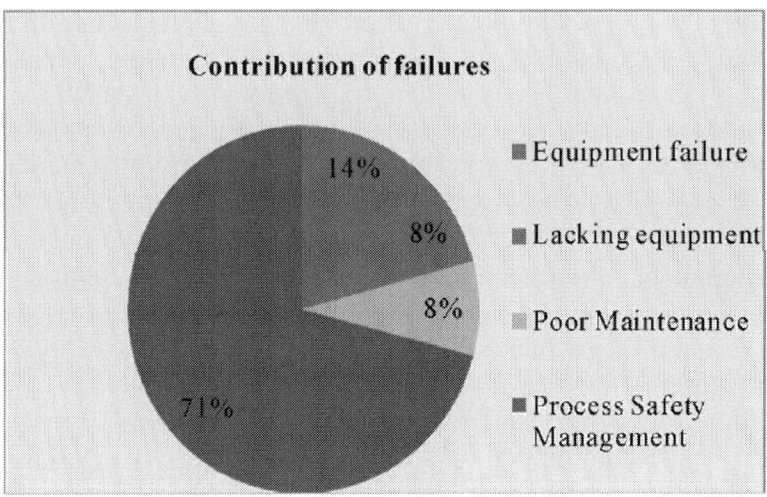

Figure 1: Contribution of failures to explosions in gas fired plant [HSE97]. "The overwhelming contributing factor that resulted in the explosions was inadequate PSM. A detailed analysis revealed that this deficient PSM was due to a lack of training, poor managerial supervision, and insufficient procedures" [HSE97: Health and Safety Executive, clause 6.2 of Contract Research Report 139/1997, "Explosions in gas fired plant" United Kingdom 1997].

The HSE [14], as part of another study, investigated 34 incidents occurred in the UK, which were the result of control system failures. This study showed that the primary causes of the control system failure were specification failures, installation and commissioning failures, failures to due changes after commissioning, design and implementation failures and operation/maintenance failures. Another major finding of the study was that the failures appeared to occur during all phases throughout the lifetime of the control system. The task of the safety management system is to prevent these failures from occurring.

Another study, in the similar field, was performed by the American Environmental Protection Agency (EPA). The EPA reviewed a large number of investigations of chemical plant accidents, over a period of several years and the EPA's Chemical Emergency Preparedness and

Prevention Office found, among other things, that operator errors were rarely the sole or even primary cause of an accident [15,16].

The majority of accidents in the process industry are not particularly the result of failure of the equipment or installation, but rather the result of inadequate safety management. Therefore, control and improvement of the safety performance should not be attempted in the area of technological improvements of the equipment, but rather in the area of safety management. The focus and attention should be to enhance the control and organization of the safety related business processes.

As mentioned earlier, the growing complexity of industrial processes has led to new kind of safety related problems. These problems concern the management and control of the safety related processes. Based on hazard investigation reports it appears that the basis of these accidents is very often the result of problems with communication and information exchange [15,16]. In other words, it can be said that the accidents occur due to the lack of adequacy of the safety framework used or improper sequence of steps evolved and safety actions taken. It can also be concluded from these studies that the safety framework used in the facilities, where accidents took place, was lacking proper verification of the safety management plan and that there were some loop holes like improper specifications, inadequate or insufficient safety measures and improper operating limits.

Hence the problem which lies in front of the process industry is to have a proper framework of safety verification which will ensure that all the inadequacies of existing safety related frameworks have been removed and that reliability should be the prime feature of such a framework. In order to incorporate any safety verification techniques in a system, it is required to have a proper framework. The use of the term verification is in line with the common definition of "verification", as answering the question "are we building the system right?" [17]. Process of verification of a new production system does not stop when production starts, but continues throughout the productive stage of its lifecycle. The basic requirements for Verification set forth in the standards are summarized as 1) Verification procedures should be performed and the results should be well documented in an auditable manner; 2) Verification should be performed by a team or personnel independent from the design and manufacturing team; 3) Verification should cover all steps in system design and manufacturing from design to final test; and 4) A Safety Verification plan should be prepared and the process of verification should be carried out on that basis [18]. Automatic and formal verification methods can guarantee that all possible situations and scenarios leading to a failure are considered in the analysis [19]. The proposed framework consists of a system of interrelation of various processes and has a set of prerequisites. These prerequisites must be clarified before the framework is incorporated and specifications should be noted. The specifications are used as guides in identifying the key behavior of the controlled process. The specifications are created from quality, operability, and safety issues that concern process engineers [20]. Before describing the proposed safety verification framework, IEC 61508 standards and the safety life cycle of a plant are explained, as illustrated in the following sections.

Safety Standards

IEC 61508 [21] published in 2000 has been adopted by many countries as their national standard and is being updated. Two significant concepts, safety life cycle and safety integrity level (SIL) [2123], appeared in IEC 61508. A necessary procedure of safety life cycle is SIL verification, which verifies whether the average probability of failure on demand (PFD avg) of designed safety related systems (SRS) meets the required failure measure. IEC 61508 is an international standard of rules applied in industry. It is titled "Functional safety of electrical/electronic/programmable electronic safety related systems". IEC 61508 is intended to be a basic functional safety standard applicable to all kinds of industry. It defines functional safety as: "part of the overall safety relating to the EUC (Equipment under Control) and the EUC control system which depends on the correct functioning of the E/E/PE safety related systems, other technology safety related systems and external risk reduction facilities."

The first premise of the standard is that there is equipment intended to provide a Function (the EUC), there is a system which controls it, and between them they pose a risk. The control system may be integrated with the EUC as, say, a microprocessor, or remote from it. The threat is shown in Figure 2 as a "risk of misdirected energy".

The standard's second premise is that "safety functions" are to be provided to reduce the risks posed by the EUC and its control system (see

Figure 2). Safety functions may be provided in one or more "protection systems" as well as within the control system itself. Any systems which are 'designated to implement the required safety functions necessary to achieve a safe state for the EUC' are classified as "safety related" systems. It is to these that the standard applies.

The standard gives guidance on good practice. It offers recommendations but does not absolve its users of responsibility for safety. Recognizing that safety cannot be based on retrospective proof but must be demonstrated in advance, and that there can never be perfect safety (zero risk), the recommendations are not restricted to technical affairs but include the planning, documentation and assessment of all activities. Thus, IEC 61508 is not a system development standard but a standard for the management of safety throughout the entire life of a system (safety life cycle), from conception to decommissioning. It brings safety management to system management and, in respect of the development of safety related systems, it brings safety engineering to software engineering.

PROPOSED SAFETY VERIFICATION FRAMEWORK

Safety analysis is a crucial part of the design and operation of chemical plants. While traditional approaches have relied heavily on qualitative analysis and expert knowledge to identify hazards, some quantitative methodologies have recently emerged [25]. As mentioned earlier, most of the LNG plants are working without clear safety frameworks. Those of them having safety features have old and obsolete frameworks. The proposed Safety Verification framework is new and acceptable to both new as well as existing plants. This framework is superior to other frameworks as it is based on the concept of safety limits rather than control limits. Involvement of safety limits extends the band of operating ranges beyond control limits which means that even if the process goes beyond the control limits, it can still be operated under constant monitoring for some more time (till it is within the safety limits). Thus, this framework delays the shutdown of a process by some time. Another very essential feature of this framework is the concept of "plant specific safety requirements". The LNG plants differ from other industrial and power plants and require a superior safety framework as they are more prone to hazardous accidents [26]. This safety framework can be considered as a dedicated LNG Plant safety Framework and employs the adequate safety measures required in the LNG plants.

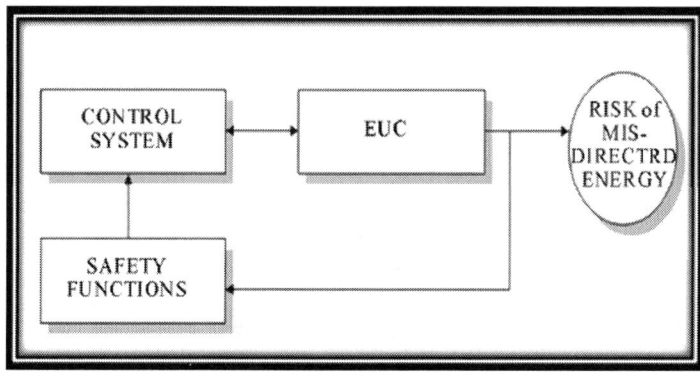

Figure 2: Control systems and safety functions for EUC.

The proposed framework is also different from the other present frameworks. While other frameworks have strict shutdown conditions, this framework provides flexibility in the shutdown of the plant. Not every abnormal condition requires a shutdown and this thought has been kept in mind while designing this framework. This feature provides additional flexibility to the safe operation of the LNG Plants. The use of an integrated network of DCS and other digital control techniques ensure that every fault causing event is taken care of and that no abnormal conditions goes unmonitored. These special features give the proposed framework, clearly an upper hand. Now we should be discussing about the framework in detail.

Activity Modeling

The proposed safety verification framework works with good effect in New Plants as well as in Existing Plants. In New Plants this framework is required to be incorporated during the Design phase of the plant while in Existing plants this framework can be incorporated by slight modification of the initial design. These changes, in the initial design, depend upon the existing level of safety in the plant and the level of safety desired. After considering these two factors the modification required in the plant design can be estimated (see Figure 3).

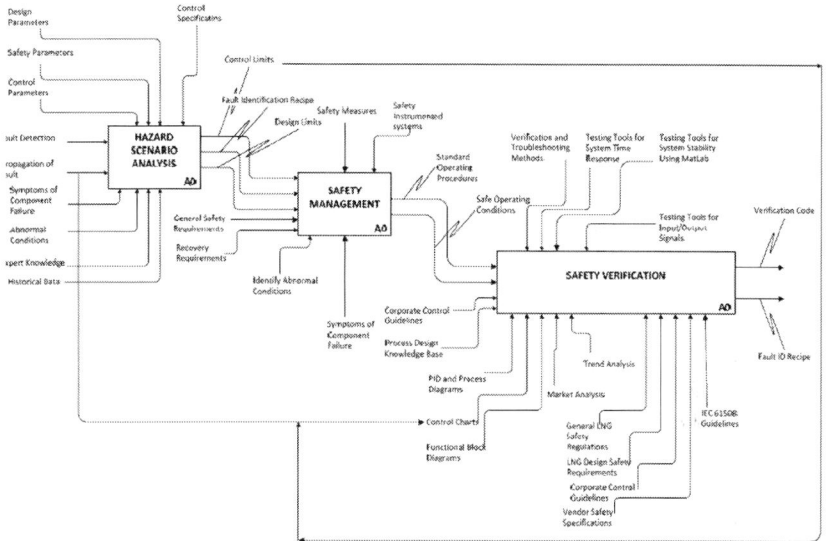

Figure 3: Safety verification framework.

As mentioned earlier, this framework consists of a system of interrelation of various processes and has a set of prerequisites. These prerequisites must be clarified before the framework is incorporated. Some of the general process prerequisites are general plant safety requirements, general recipe for recovery, symptoms of failure mode etc. The first process is the hazard scenario analysis and then, the second process is to have a safety management plan for the safe operation of LNG facility. Then keeping in view the general safety requirements of the plant, general recipe of recovery and failure of mode, we verify and the safety requirements. The third process is the verification of the safety management plan once the safety requirements are chalked out. This process of verification is to verify the complete safe operation of the plant according to the general LNG Safety Regulations and LNG Design Safety Requirements Guidelines. The complete framework and all of its processes and sub processes are designed to work in accordance with IEC 61508. It is a generic international standard entitled to achieve safety of the system, as mentioned in section 2.3 of the paper. In order to understand the framework, it is essential to understand its processes and sub processes which can be broadly classified as Hazard Scenario Analysis, Safety Management, and Verification (and Testing). These are described in more details in the following sections.

Hazard Scenario Analysis

Hazard Scenario Analysis is the most basic and fundamental block of any safety related framework (see Figure 4). Without proper identification of a hazard scenario, we cannot control the operation of any process in a plant. Also, without it, talking about safety or safe operation would be baseless. Unless and until the hazard scenarios are analyzed, one cannot determine the ranges in which a particular equipment or process should operate, and the ranges beyond which a particular process or equipment is uncontrollable and unsafe to operate [27]. From this discussion, we can conclude that limits estimation is an integral part of hazard scenario analysis and further we can conclude that hazard scenario analysis and then determining the limits forms the first block of activity modeling for any framework.

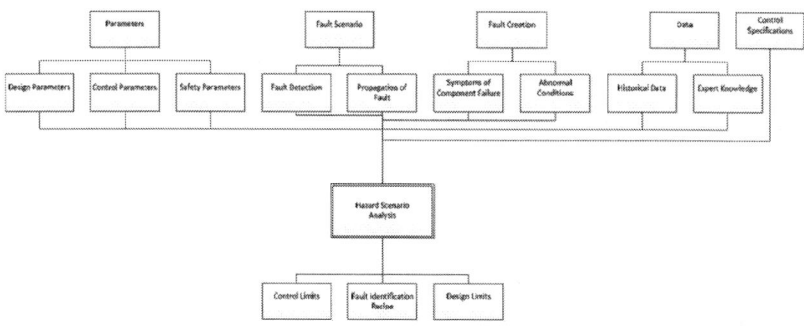

Figure 4: Pictorial representation of hazard scenario analysis block of the framework.

In order to estimate the limits, we require the process parameters, variables and units. Process parameters such as design parameters, control parameters and safety parameters are essential to be known before limit estimation. Variables needed to be known are the process variables and control variables. Similarly, process units and functional units of a process are required. Another very important thing which should be placed at desk before calculating the limits is the historical data of the process. With this data, we come to know about the behavior of the process in past and we can make changes to our calculations accordingly. Also some specifications, known as Control Specifications, should be known as a process is required to operate within these specifications.

With all the above things at hand, viz. the parameters, the historical data and the units, the variables and the control specifications, one determines the limits of safe operation and identifies the unsafe zones while an equipment or process is in operation. Along with the limits estimation, we are keenly interested in the propagation of a fault. If the propagation of a fault is closely monitored, the fault itself can be suppressed in its initial stages. Events like component failure and abnormal conditions also lead to fault propagation. Thus fault detection, as early as possible, acts as a useful tool in analyzing the hazard scenario.

The analysis of hazard scenario means calculating the control limits (the limits of operation within which the process is safe and controllable and is most desired to work), the safety limits (the limits beyond the control limits domain, where the process is uncontrollable but safe to operate for a short time before it can be restored back to the control limits domain) and the design limits. Also a fault id recipe is generated. These three limits together with the fault id recipe, when determined and estimated, form the input for the safety management plan, which is the second block of the framework.

Safety Management

Safety Management is the second block of the framework (see Figure 5). This more of a plan than a block which is required to manage all the essential safety needs for any plant in general including the LNG plants. This plan deals with the procedures of establishing the safety requirements and modes of failure prevention for a plant. In order to have such a plan, the most important prerequisites are the safety requirements, the limits of operation and the modes of preventing failure [28].

In order to comprehend the plan, we must, at the beginning, be familiar with the safety requirements. These safety requirements are plant specific. For instance, an LNG plant may have a different set of safety requirements than a nuclear power plant or a thermal power plant. To have these plant specific safety requirements we must know the general safety requirements and the recovery requirements. The general safety requirements are the requirements which are needed in the normal operation of a plant whereas the recovery requirements are needed, in case, when the process conditions remain no longer safe and a recovery to the safe mode is required. These are "backup requirements", but are important from the perspective of safe operation of a plant. Then we need the limits, whose estimation we have already discussed in the previous section. Operating a plant in safe mode means operating it within these predetermined limits, regular monitoring the process parameters and taking necessary recovery actions when needed.

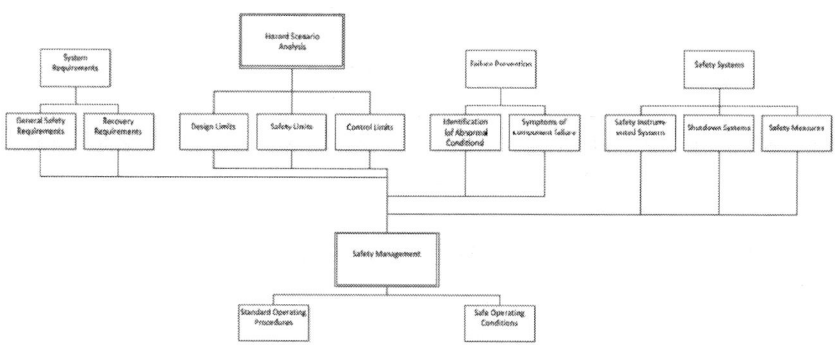

Figure 5: Pictorial representation of the safety management block of the framework.

Next important thing needed for a safety management plan are the modes of failure prevention. Just by incorporating the recovery requirements whenever a plant goes into the unsafe zone, does not solve the purpose. In fact, incorporating the recovery requirements should be the last step, before shut down, whereas the failure prevention modes must be running when the plant is operating even at normal conditions. This is to ensure that a plant operates at in the safe zone and a need to incorporate recovery requirements must not arrive. These include complete constant monitoring of the abnormal conditions and the symptoms of component failure. Once an abnormal condition is identified, it must be indicated to the operator, who must take the necessary actions to maintain normalcy again. It is worth making note of that not all the abnormal conditions lead to system failure. So it must be identified whether an abnormal condition would lead to a system failure or not, from the past experiences, and take necessary corrective measures accordingly. This is the most decisive step in order to prevent accidents in any industrial plant. As we know, the slightest of risk may lead to a hazard; therefore past experiences should be taken into account only if the operator is surely certain.

The last, but not the least, prerequisite are the safety systems which include the safety integrated systems, shutdown systems and other similar systems which are designed for the last step to be taken, in maintaining the normalcy of the plant. Once we have the above mentioned units, we can say that the safety management plan is comprehended correctly and our plant is safe to operate.

Safety Verification

Verification and testing forms the third block of the proposed framework (Figure 6). No safety management strategy is trustworthy unless verified. Thus a good safety management plan is one which can be duly verified and tested in various different situations. Thus Verification and Testing can be regarded as the most important block of the framework.

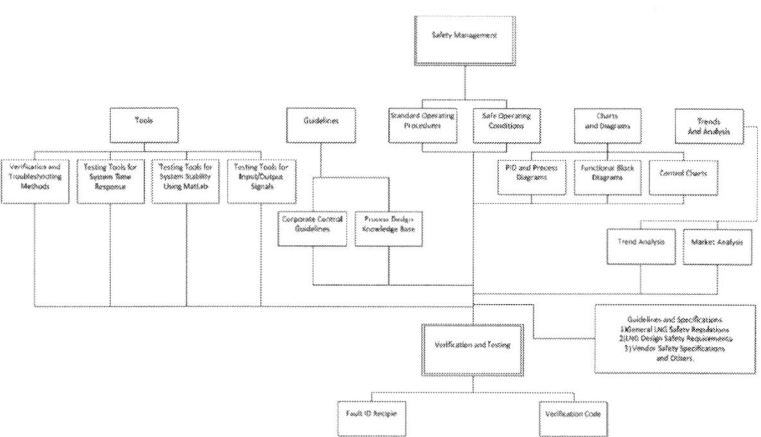

Figure 6: Pictorial representation of the verification and testing block of the framework.

To properly verify a safety scenario of a plant we require certain tools, guidelines and trends of performance (of the process/equipment or parameters). Tools are the techniques used for proper verification. These can be the verification and troubleshooting methods, tools for testing the time domain and frequency domain response of a particular process or a group of processes, as desired, tools for testing the stability using known methods like bode plot, nyquisyt plot, etc., using MATLAB and tools for testing input output signals. The tools can be operated on various platforms like MATLAB, SIMULINK, MAPLE SIM, etc. for testing purposes.

We also need to verify some standard operating procedures and safe operating conditions. For these we need a set of guidelines which can be corporate control guidelines or those of the process design knowledge base. Certain charts and diagrams like the P&ID and process diagrams, FBD (functional block diagram) and control charts are also helpful during the verification phase.

Another important necessity is the availability of trends for various parameters and process variables. These are the behavior of the parameters with respect to time in a certain given conditions. These can be plotted and analyzed for detailed understanding of the trends which they follow. It is an important aspect of safety verification as these provide the inside knowledge of the things happening in a process. Analyzing the market trends is also a good practice during verification.

Thus to summarize, the verification block includes the verification of safety measures and makes sure that the readings obtained after the verification of safety procedures are valid as per the standards set by the industry. There are many regulations, requirements, guidelines and specifications which must be verified before deeming any plant safe. The most common ones which must always be verified are General LNG Safety Regulations, LNG Design Safety Regulations, Corporate Control Guidelines, IEC 61508 Guidelines, IEC 61511 Guidelines, ISAS84.01 Guidelines and others. The verification code is generated at the end of the verification phase. Once we have studied the framework, we need to identify a hazard scenario for proper case study and mapping of the hazard scenario to the safety and verification framework proposed above. We need to obtain data so that we can study trends occurring during our case study. The next section deals with the case study, results and discussions (see Figure 7).

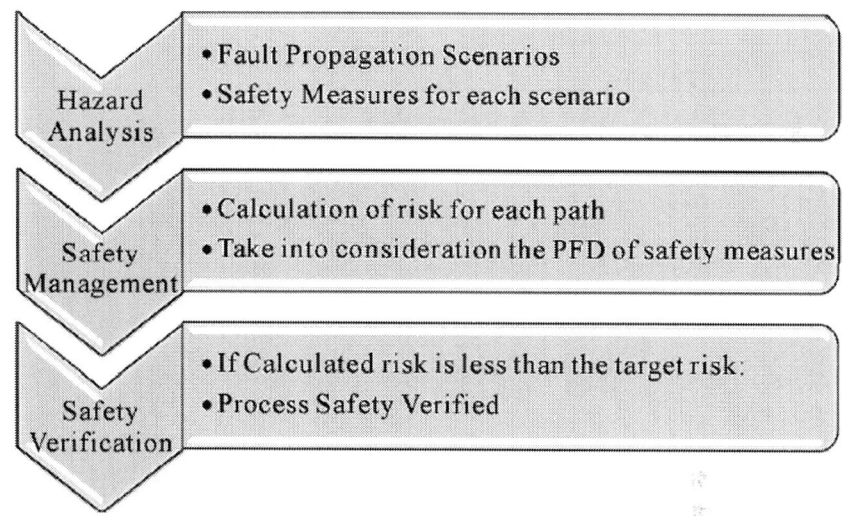

Figure 7: Flow of process as per the safety verification framework.

DETAILED SAFETY VERIFICATION ALGORITHM

A Flowchart Algorithm for the proposed framework is shown below (Figure 8 and 9):

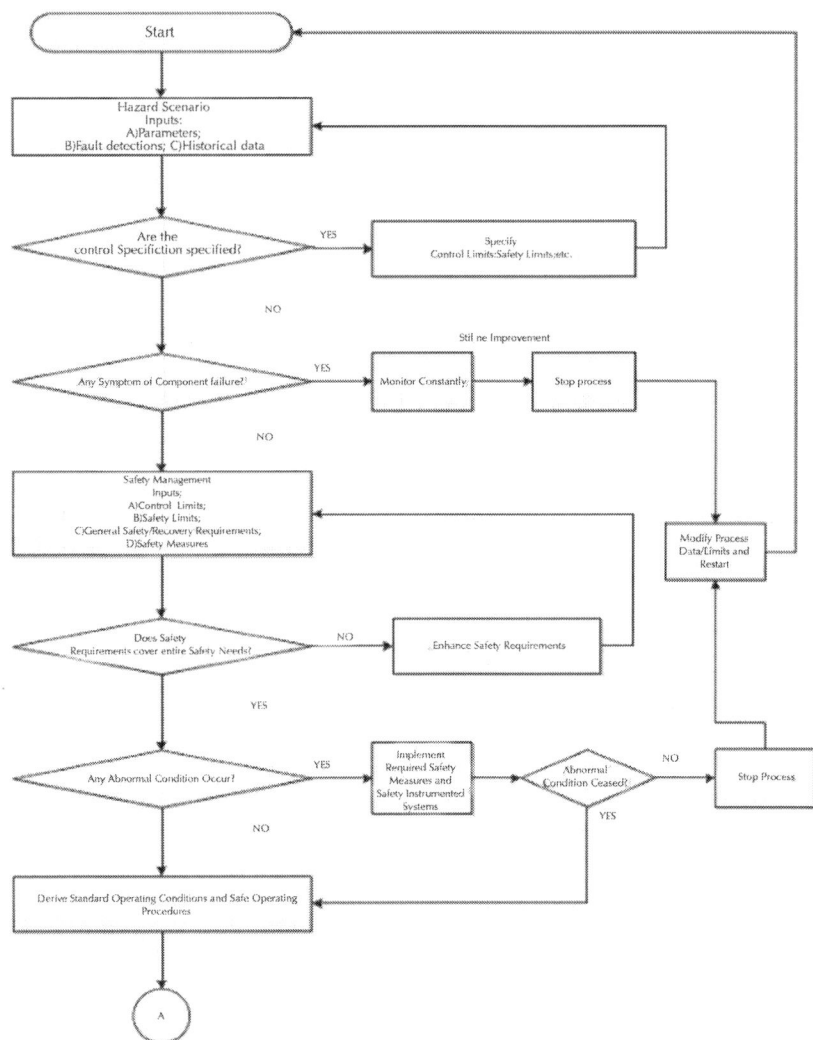

Figure 8: Safety verification algorithm (Part1).

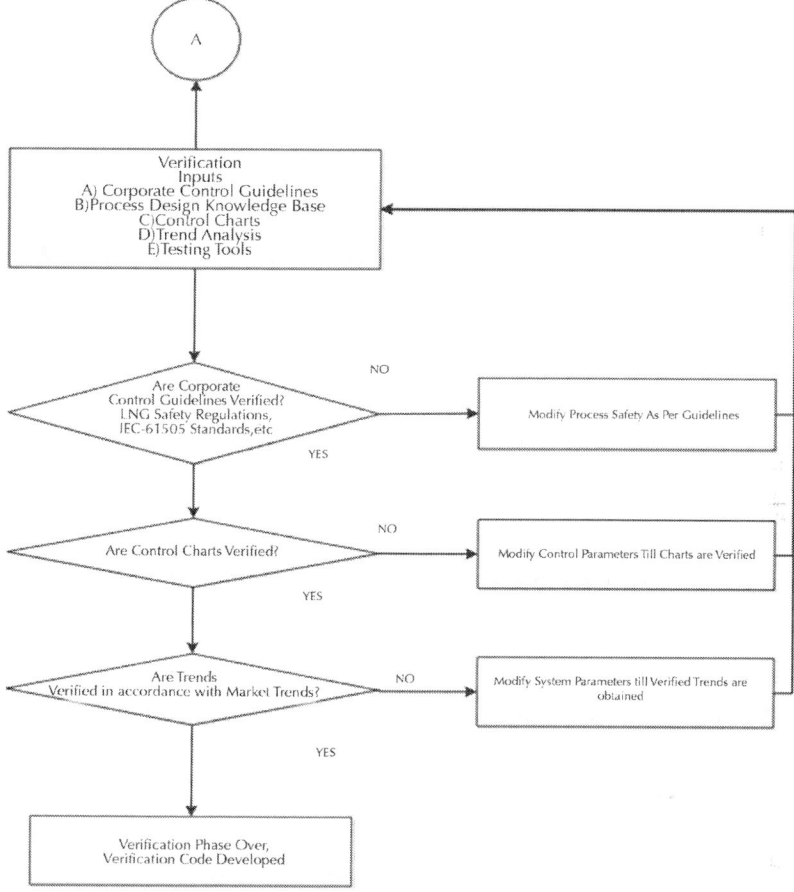

Figure 9: Safety verification algorithm (Part2).

SOLUTION IMPLEMENTATION

Case Study

In order to illustrate the proposed idea, a case study proposed using experimental plant called GPlant, which was developed in IGPS group in Okayama University as part of industrial collaboration project in Okayama, Japan [29]. GPlant is an experimental plant that consists of two stainless steel tanks, one with heaters to increase the temperature of the water to a predefined set point. DCS Centum CS3000 from Yokogawa is installed [30]. The P & ID of the constructed experimental plant is shown in Figure 10. Cold water is circulated from the tank TANK2 to the heat exchanger HEX1 and then back to the tank TANK2. Similarly, hot water is circulated from tank TANK3 to heat exchanger HEX1 and then back to tank TANK3. Hot water is used to heat cold water in tank TANK2 where temperature increase is monitored in real time basis within DCS. Similarly, other process variables (sensors) are monitored within DCS for process control and safety. Flow rate of the cold water circulation is controlled using control valve CV3. Heat exchanger level is monitored to avoid overflow. Levels in TANK1 and TANK2 are monitored to avoid overflow. Temperature in TANK2 is controlled to avoid overheating. Alarms are defined for all critical set points in GPlant. For example, alarm is generated when temperature in TANK2 exceeds a predefined set point. The experimental plant is used to simulate and diagnose process faults. For example and in order to simulate leak in heat exchanger HEX1, downstream valve is slightly opened during the circulation of cold water. Readings are obtained for four process variables: TC1 (temperature in the cold water circulation loop), TC2 (temperature in the inlet of hot water), TK2 (temperature in tank TANK2) and TK3 (temperature in tank TANK3).

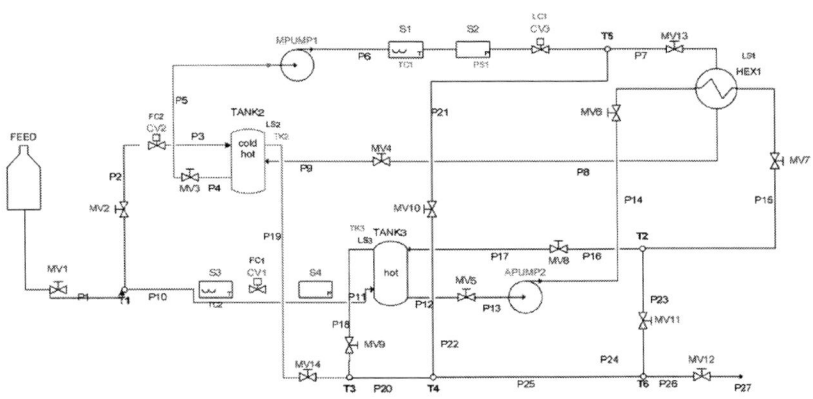

Figure 10: P & ID of GPlant (Gabbar, 2007).

For Hazard Analysis we take a scenario in which there is a high flow of liquid in the TANK2 (shown by the red bold lines in the P&ID)

which eventually leads to overflow. This high flow of fluid may cause vibrations in the tank and also offer some blockage to the outflow of the fluid. A detailed cause effect study and the propagation of fault leading to a hazard, is shown in the Figure 11. Primary causes, such as high/low temperature, high/low flow, overflow, impurities, etc. lead to the initiation of the hazard. They have a Low Qualitative Hazard Magnitude (QHM) as the probability of their occurring in any process is high and the probabilistic risk associated with them is quite low. Though the QHM associated with them is low, they cannot be neglected as they lead to the initiation of a hazard. Strong monitoring is needed and proper action (implementation of safety measures) should be taken depending upon the behavior of these parameters. Primary causes lead to primary events, which may be vibrations in the tank or blockage due to uneven flow in this case. These primary events form the secondary causes of the fault propagation. These secondary causes have a medium QHM and a high probabilistic risk associated with them. These secondary causes lead to secondary events or tertiary causes, which may be corrosion of the tank material. Tertiary causes lead to tertiary events (or quaternary/fourth degree causes) like leak or reduced mechanical strength. The fourth degree causes the most dangerous ones with an extremely high QHM and a very large probabilistic risk associated with them. These eventually lead to hazard which may be fire, intoxication of air or explosion in this case. Thus we should implement appropriate safety measures at each level of fault propagation (Figure 12).

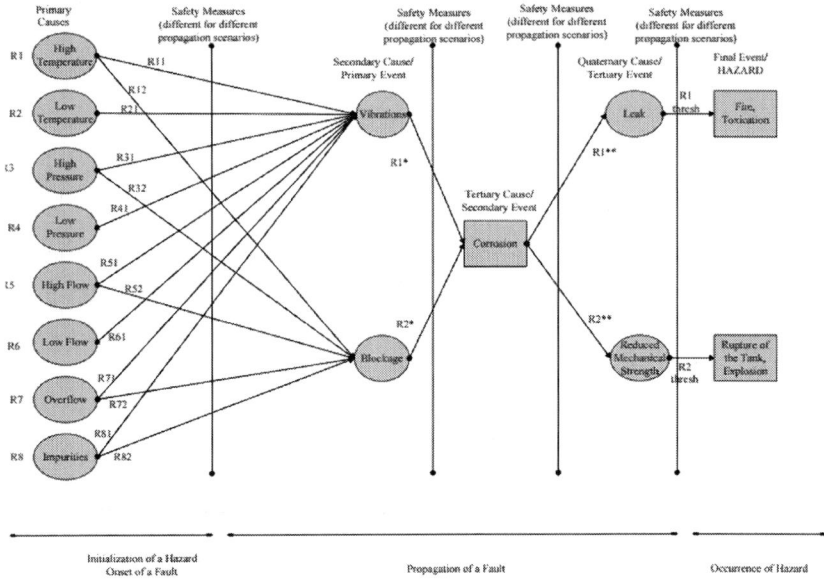

Figure 11: Fault propagation and intermediate causes and effects.

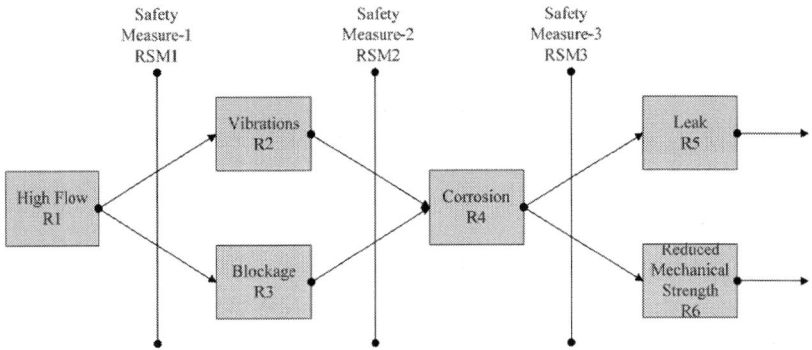

Figure 12: Propagation of a fault for a particular event.

Table 2: Failure rates [31]

Risk	Meaning	Failure Rate(per year)
R1	Risk associated with High Flow	10
R2	Risk associated with vibrations	2
R3	Risk associated with Blockage	1.1
R4	Risk associated with Corrosion	0.9
R5	Risk associated with Leak	0.06
R6	Risk associated with reduced Mechanical Strength	0.09
RSM1	Risk associated with Failure of Safety Measure1	0.003
RSM2	Risk associated with Failure of Safety Measure2	0.003
RSM3	Risk associated with Failure of Safety Measure3	0.003

Quantitative Hazard Analysis

Let us assume that that initializing event leading to a hazard is High Flow in the Tank (Figure 12). If the Safety Measure1 employed to check the flow rate of the tank fails, this high flow will lead to Vibrations and/or Blockage. Again if Safety Measure2 fails to perform its task, these Vibrations and Blockage may cause Corrosion. And if Safety Measure3 also fails, this Corrosion may lead to Leak or Reduced Mechanical Strength which may lead to fire, intoxication or even explosion of the tank. This is how a fault propagates and ultimately leads to a hazard.

Risk associated with Safety Measure is directly related to the Probability of Failure on Demand (PFD) of that safety Measure. Now our aim is to find out whether our system is safe or not. For this we will take individual fault propagation events into consideration and calculate the total risk associated. This "total risk associated" is the magnitude of risk which will lead and onset of a fault of the hazard.

Analysis of Individual Fault Propagation Events

Let us assume that the magnitude of failure be a constant. This magnitude of failure is actually given by the company based on the historical data of accident and the consequences occurred per event. We are assuming it to be a constant because it is a number which can be later substituted to get more correct information. Thus assuming magnitude of failure to be a constant, we can now say that the risk associated with any event is directly proportional to its failure rate and is a function of failure rate.

Risk Associated = f (failure rate)

The risk associated with fault propagation path1 (Figure 13) is calculated as below:

Risk Associated (Path1) = Rl*RSMI*R2*RSM2* R4*RSM3 *R5

Risk Associated (Path1) = 10 x 0.003 x 2 x 0.003 x 0.9 x 0.003 x 0.06 = 2.916E8

The risk associated with fault propagation path2 (Figure 14) is calculated as below:

Risk Associated (Path2) = Rl*RSM1*R2*RSM2* R4*RSM3 *R6

Risk Associated (Path2) = 10 x 0.003 x 2 x 0.003 x 0.9 x 0.003 x 0.09 = 4.378E8

The risk associated with fault propagation path3 (Figure 15) is calculated as below:

Risk Associated (Path3) = Rl*RSMI*R3*RSM2* R4*RSM3 *R5

Risk Associated (Path3) = 10 x 0.003 x 1.1 x 0.003 x 0.9 x 0.003 x 0.06 = 1.604E8

The risk associated with fault propagation path4 (Figure 16) is calculated as below:

Risk Associated (Path4) = R1 *RSM 1 *R3*RSM2* R4*RSM3*R6

Risk Associated (Path4) = 10 x 0.003 x 1.1 x 0.003 x 0.9 x 0.003 x 0.09 = 2.406E8

Calculation of Total Risk Associated (TRA)

Now the Total Risk Associated (combined of all paths) that an onset of a fault, i.e. high flow, will lead to a hazard i.e. fire or explosion, is the sum total of the total risk associated of all the paths (see Table 3).

Now if the total risk associated is less than the threshold risk (level of acceptable risk), then our process is safe, otherwise it is not. This threshold risk is calculated from the process historical data and other equipment data. It is calculated on the basis of the following formula:

Threshold Risk (TR) = Frequency of Failure*Magnitude of failure

Again assuming the risk as a function of failure rate, we can calculate the threshold risk. The typical value of failure rate can be taken as per year [31]. This if the TRA is more than this value, our process is unsafe (Table 4).

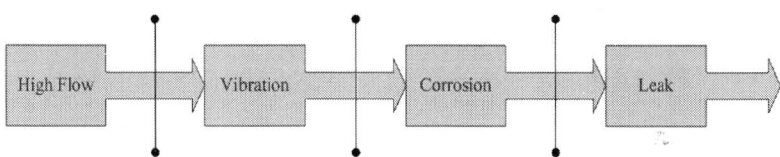

Figure 13: Individual fault propagation event (Path1).

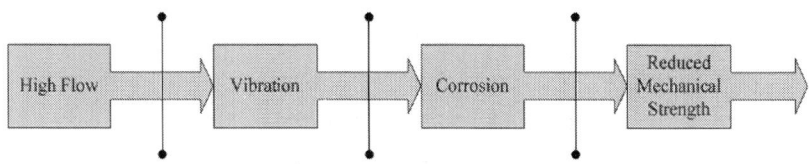

Figure 14: Individual fault propagation event (Path2).

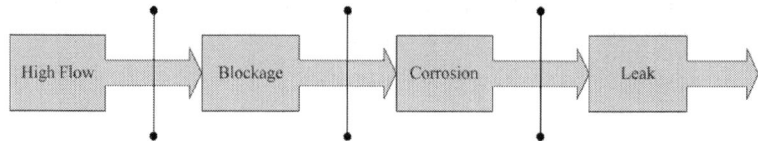

Figure 15: Individual fault propagation event (Path3).

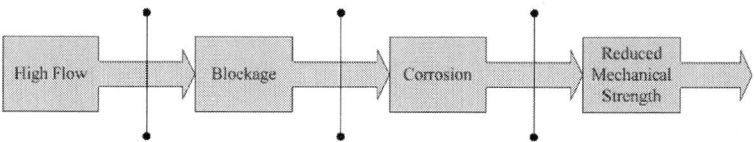

Figure 16: Individual fault propagation event (Path4).

Table 3: Calculation of total risk associated

Total Risk Associated(TRA) = Risk Associated (Path1) + Risk Associated (Path2) + Risk Associated (Path3) + Risk Associated (Path4)
Total Risk Associated(TRA) = 2.916E8+4.378E8+1.604E8+2.406E8 = 1.1304E7

Table 4: Verification of safety

VERIFICATION
TRA=1.1304E7
TR=5E6
TRA<TR;PROCESS SAFE
SAFETY VERIFIED

CONCLUSIONS

The proposed safety verification framework is indeed very necessary in order to have a safe and a fail proof safety plan for any LNG plant. It is new and acceptable to both new as well as existing plants. It is flexible in the sense that it can be applied to both new and existing plants with same effect. As we know that we cannot ignore safety concerns in any LNG plant, we can conclude that safe operating conditions are of huge importance in any LNG facility. This safety framework operates on the concepts of safety limits and therefore provides an extended range of safe operation. The proposed framework is also different from the other present frameworks. While other frameworks have strict shutdown conditions, this framework provides flexibility in the shutdown of the plant. Not every abnormal condition requires a shutdown and this thought has been kept in mind while designing this framework. This feature provides additional flexibility to the safe operation of the LNG Plants. The use of an integrated network of DCS and other digital control techniques ensure that every fault causing event is taken care of and that no abnormal conditions goes unmonitored. These special features give the proposed framework, clearly an upper hand. At the last, it can be said that this safety framework can be considered as a dedicated Process Industry Safety Framework and employs the adequate safety measures required in the LNG plants.

REFERENCES

1. R. Ali, "Safety Life Cycle—Implementation Benefits and Impact on Field Devices," ISAExpo 2005, Chicago, 2527 October 2005.
2. G. Holger and S. T. Henner, "Process Hazard Identification during Plant Design by Qualitative Modelling, Simulation and Analysis," European Symposium on Computer Aided Process Engineering, Vol. 23, Supplement 1, 1999, pp. S59S62.
3. H. A. Gabbar, "Integrated Framework for Safety Control Design of Nuclear Power Plants," Nuclear Engineering and Design, Vol. 240, No. 10, 2010, pp. 35503558. doi:10.1016/j.nucengdes.2010.07.024
4. IEC 61508, Functional Safety of Electrical/Electronic/Programmable Electronic SafetyRelated Systems, 1998/ 2000.
5. IEC 61511, Functional Safety: SafetyInstrumented Systems for the Process Industry Sector, Draft version 1999.
6. ANSI/ISA S84.01, Research Triangle Park, 1996

7. B. Knegtering, "The Impact of IEC 61508 and IEC 61511 on Dutch Industry Epigram," Official Journal of Core Interest User, Group of Programmable Electronic Systems, London, Autumn 2000, unpublished.
8. B. Knegtering, "Safety Lifecycle Management," Automation in Petro Chemicals Industry Conference, University of Ontario Institute of Technology, 2000 Simcoe St. N, Oshawa, Canada.
9. F. P. Lees, "Loss prevention in the process industries," 2nd, Edition, ButterworthHeinemann, Oxford, 1996.
10. B. Knegtering, "Application of Micro Markov Models for Quantitative Safety Assessment to Determine Safety Integrity Levels," ISAExpo, Houston, 1923 October 1998.
11. B. Knegtering and A. C. Brombacher, "A Method to Prevent Excessive Numbers of Markov States in Markov Models for Quantitative Safety and Reliability," ISATransactions, Vol. 39, No. 3, 2000, pp. 363369. doi:10.1016/S00190578(99)000415
12. Health and Safety Executive, "Explosions in gasfired plant," Clause 6.2 of Contract Research Report 139/1997, UK, 1997.
13. Bradley, "The Reliability Challenge," Presentation handouts Conference, London, 1999.
14. Health and Safety Executive, "Out of Control HSE Books," United Kingdom 1995.
15. B. Felton, "Safety study IDs Leading Causes of Accidents," InTech, Morn Hill, 2001, p. 77.
16. J. Belke, "Chemical Accident Risks in US Industry—A Preliminary Analysis of Accident Risk Data," US Hazardous Chemical Facilities EPA, September 2000.
17. M. H. C. Everdij, H. A. P. Blom, J. J. Scholte, J. W. Nollet and B. Kraan, "Developing a Framework for Safety Validation of MultiStakeholder Changes in Air Transport Operations," Safety Science, Vol. 47, No. 3, 2009, pp. 405420.
18. A Fukumoto, T. Hayashi, H. Nishikawa, H. Sakamoto, T. Tomizawa and T. Yokomura, "A Verification and Validation Method and Its Application to Digital Safety Systems in ABWR Nuclear Power Plants," Nuclear Engineering and Design, Vol. 183, No. 12, 1998, pp. 117132.

19. S. H. Yang, L. S. Tan and C. H. He, "Automatic Verification of Safety Interlock Systems for Industrial Processes," Journal of Loss Prevention in the Process Industries, Vol. 14, No. 5, 2001, pp. 379386. doi:10.1016/S09504230(01)000146
20. S. Brown, "Overview of IEC 61508: Functional Safety of Electrical/ Electronic/Programmable Electronic SafetyRe lated Systems," Computing and Control Engineering Jour nal, Vol. 11, 2000, p. 11.
21. P. Stavrianidis and K. Bhimavarapu, "PerformanceBased Standards: Safety Instrumented Functions and Safety Integrity Levels," Journal of Hazardous Materials, Vol. 71, No. 13, 2000, pp. 449465.
22. IEC 61508, Functional Safety of Electrical/Electro nic/ Programmable Electronic SafetyRelated Systems, International Electro Technical Commission, Reference: IEC 615083 ed 2.0.
23. F. Redmill, "An Introduction to the Safety Standard IEC 61508," Journal of the System Safety Society, Vol. 35, No. 1, 1999, pp. 2125.
24. C. S. Adjiman, "Safety Verification in Chemical Plants: A New Quantitative Approach," Computers & Chemical Engineering, Vo. 23, Supplement 1, 1999, pp. S581S584. doi:10.1016/ S00981354(99)801434
25. H. A. Gabbar and P. Sauer, "Knowledgebase and Acquisition System for Failure and Accident Analysis of Gas Processing Facilities," International Workshop on Real Time Measurement, Instrumentation & Control, Oshawa, 2526 June, 2010.
26. H. A. Gabbar and R. Bedard, "Hazard Analysis and Accident Prediction for LNG Plants," International Workshop on Real Time Measurement, Instrumentation & Control, Oshawa, 2526 June, 2010.
27. Y. Shimada and T. Kitajima, "Framework for Safety Management Activity to Realize OSHA/PSM," International Workshop on Real Time Measurement, Instrumentation & Control, Oshawa, 2526 June, 2010.
28. H. A. Gabbar, H. E. Sayed, A. S. Osunleke and H. Masanobu, "Analytical Process and System Design of Integrated Fault Diagnostic System," International Journal of Process Systems Engineering, Vol. 1, No. 1, 2009, pp. 6681.

29. E. Nasimi and H. A. Gabbar, "Development of Support Tool for Control Design of Nuclear Power Plant Using Hierarchical Control Chart (HCC)," Journal of Process Systems Engineering, Vol. 1, No. 2, 2010, pp. 150168.
30. H. A. Gabbar, H. E. Sayed, A. S. Osunleke and H. Masanobu, "Design of Fault Simulator," Journal of Reliability Engineering and System Safety, Vol. 94, No. 8, 2009, pp. 12891298. doi:10.1016/j.ress.2009.01.006
31. A Blanchard, "Savannah River Site Generic Data Base Development," Westinghouse Savannah River Company, Aiken, NTIS Order No. 29808.

# Chapter 8

Characterisation and Pre-concentration of Chromite Values from Plant Tailings Using Floatex Density Separator

C. Raghu Kumar[1], Sunil Tripathy[1], and D.S. Rao[2]

[1]R&D Department, TATA Steel Limited, Jamshedpur, India
[2]Mineralogy Dept., IMMT, Bhubaneswar, India

ABSTRACT

Classification is a method of separation of fines from coarse particles and also lighter particles from heavier particles. The conventional classifiers, such as, hydrocyclone or mechanical classifiers, decreases the efficiency of the grinding and concentration circuits due to their imperfect separation. In the process of improving the efficiency of classification, a device that has been gaining popularity in recent years is the teeter-bed or hindered-bed separator such as Floatex density separator. Generally for processing chromite ores, different types of gravity methods are employed after crushing, grinding followed by classification. The Tata Steel Chrome Ore Beneficiation (COB) plant is generating 50 tph of tailings assaying 17% Cr_2O_3. A critical review on practice of the plant operating personnel is concerned in the grade-recovery characteristics of unit operations. But separation insight and influence of different operating and process parameters are essential to understand and control the process. The objective of the present investigation was to study the effect of the important operating variables on floatex density separator and preconcentration of COB plant tailings for the further beneficiation process and found that significant removal of iron bearing mineral such as goethite and silica is possible using FDS in a single stage operation. The maximum of 83% recovery of chromite is possible with 22 to 23% Cr_2O_3 content and thus obtained FDS underflow is suitable for flotation circuit. A low teeter water flow rate with a high bed pressure removes iron bearing mineral like goethite efficiently in an FDS.

INTRODUCTION

Classification is a method of separation of fines from coarse particles and also lighter particles from heavier particles. This is performed on the basis of the velocity with which the grains fall through a fluid medium generally water or air [1]. In view of the fact, that the velocity of particles in a fluid medium is dependent not only on the size, but also on the specific gravity and shape of the particles. The conventional classifiers, such as, hydro cyclones or mechanical classifiers, decreases the efficiency of the grinding and concentration circuits due to their imperfect separation. Several attempts have been made to improve the efficiency of classification. They include the use of screens instead of classifiers [2], the use of cone classifiers to process hydro cyclone under flow [3,4] and two stage classification by hydro cyclones [5,6,7].

In the process of improving the efficiency of classification, a device that has been gaining popularity in recent years is the teeter-bed or hindered-bed separator such as Floatex density separator. The upward flow of elutriation water creates a fluidized "teeter-bed" of suspended particles. The small interstices within the bed create high interstitial liquid velocities that resist the penetration of the slow settling particles.

As a result, small/light particles accumulate in the upper section of the separator and are eventually carried over the top of the device into a collection launder. Large/heavy particles, which settle at a rate faster than the upward current of rising water, finally pass through the fluidized bed and are discharged out through the bottom of the separator as underflow.

The major Indian chromite deposits are located at Sukinda region of Orissa state. The depletion of high grade ore resources and having a variety of gangue minerals such as goethite, serpentine, olivine and talc, etc., has lead to the utilisation of lean ore after beneficiation. Further, ample amount of plant tailing generation have gained importance from the economics, conservation and ecology point of view [8].

Generally for processing chromite ores, different types of gravity methods are employed after crushing, grinding followed by classification. The Tata Steel Chrome Ore Beneficiation (COB) plant, designed to produce concentrate of +46% Cr_2O_3 with a recovery of 70% from a feed of 30- 35% average Cr_2O_3. Presently the COB plant produces 50 tph of tailings analysing 17% Cr_2O3, which is high [9]. This suggested for the incorporation of additional circuit comprising of hydro cyclone for pre-concentration by de-sliming and multigravity separator and wet high intensity magnetic separator for the upgradation up to the required quality. For recovering chrome values from the Karagedik Concentrate tailings a circuit comprising of wet high intensity magnetic separator and column flotation for producing a concentrate assaying 46 to 48% Cr_2O_3 was studied by Guney et. al [10].

A critical review on practice of the plant operating personnel is concerned in the grade-recovery characteristics of unit operations. But separation insight and influence of different operating and process parameters are essential to understand and control the process. The objective of the present investigation was to study the effect of the important operating variables on floatex density separator and pre-concentration for the further beneficiation process.

EXPERIMENTAL

The Sukinda COB Plant tailing sample of Tata Steel, India was the feed material in the present studies. The plant tailing as received sample was subjected to characterization in terms of size and chemical assay, size wise microscopic liberation studies. The mineralogical studies were carried out with the help of stereomicroscope and reflected light microscope.

The experimental campaign was undertaken in a lab scale Floatex density separator (Model No. LPF-0230), supplied by Outokumpu of 230 mm X 230 mm cross section and 530 mm high (square tank height) followed by a 200mm high conical section. The Floatex density separator (FDS) can be divided into three main zones,

- the upper zone (zone A) above the feed inlet,
- the intermediate zone (zone B) between the feed inlet, and teeter water addition point, and
- the lower section (zone C) below the teeter water addition point.

Feed slurry is introduced to the FDS tangentially through a centralized feed well that extends to approximately one third of the main tank length. Fluidizing (teeter water) is introduced over the entire cross-sectional area at the base of the teeter chamber through evenly spaced water distribution pipes. As the feed enters the main separation zone it expands into a teetered or fluidized bed as a result of the rising current of water. The teeter water flow rate is dependent upon a) feed particle size distribution, b) density and c) the desired cut-point for the separation.

The separation takes place in zone B and the separated lighter/finer particles and the coarser/heavier particles leave the separator through zone A and zone C respectively. This separator is equipped with a pressure sensor mounted in zone B above the teeter water pipes and an underflow discharge control valve. The pressure, sensed by a level sensor, is transmitted to the underflow control valve using a specific-gravity set-point controller. The instrumentation helps in maintaining a constant height of the teeter bed and a steady discharge of the underflow. FDS is an efficient hydraulic classifier for classify the material based on their slip velocity. The slip velocity is the relative velocity between the particles and the water velocity and is the function of size and density of the minerals [11]. A schematic diagram of Floatex density separator is presented in Figure 1.

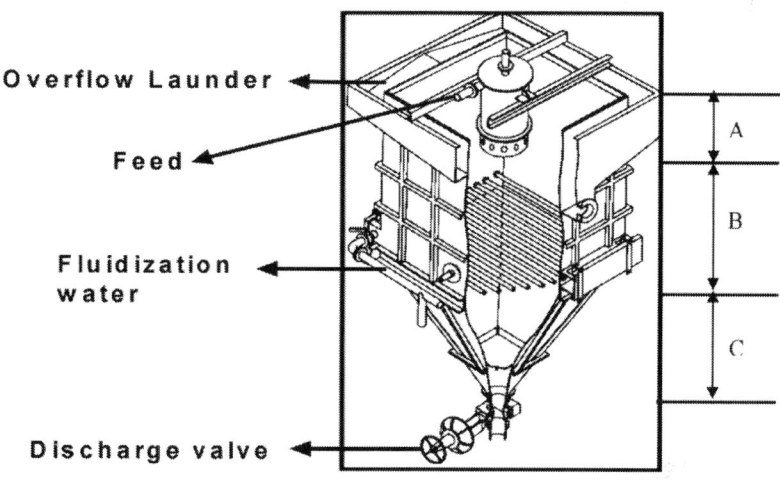

Figure 1: Floatex density separator.

About 20 tests were conducted with different combination of operating variables which is shown in table 1. Both the underflow and overflow products for each experiment were collected, dried, weighed and subjected to granulometry and chemical analysis. The experimental data were scrutinized and the performance of the FDS was quantified in terms of cut size (D_{50}), Imperfection (I) and the Cr_2O_3 percentage recovery of the underflow in each condition. The effect of teeter water flow rate and bed pressure was evaluated.

Table 1: Design of tests with floatex density separator

Variables	Level			
	1	2	3	4
Teeter water flow rate (in lpm)	6	8	10	12
Bed pressure (in bar)	0.06	0.065	0.07	0.075

RESULT AND DISCUSSION

Characterisation Studies

As received sample contains 17.76% of Cr_2O_3 and the major impurities are Fe(T) 22.28%, Al_2O_3 22.40%, SiO_2 5.71%, MgO 3.39%, CaO 0.17% and LOI of 12.55%. The chemical analysis of each size fraction was carried out in an ICP analyser and the analysis data is shown in Table 2. It may be seen that the 250 micron size fraction contained 32.5% by weight and assayed 10.3% Cr_2O_3. Whereas less than 25 micron size fraction contains 11.67% Cr_2O_3 with maximum Fe(T) i.e 33.16%. It can also be seen from the table that the Cr_2O_3 content increasing as size decreases. Where as in the case of iron there is no much variation upto plus 25 microns size. From the Figure 2 it is evident that 80% of the particle size below 410 microns whereas 50% of the sample is below 195 microns. It has been observed that particles below 25 microns size are 19.52% by weight.

Table 2: Size analysis and size wise chemical analysis of the as received sample

Mesh size (micron)	Wt% Retained	Assay Value (%)						
		Cr_2O_3	Fe(T)	Al_2O_3	SIO_2	CaO	MgO	LOI
+500	11.43	8.81	22.46	30.22	6.46	0.17	1.55	18.51
+250	21.10	11.46	20.61	27.76	5.21	0.10	2.06	15.64
+150	22.26	24.95	19.01	23.20	4.64	0.18	4.93	11.28
+104	9.48	28.75	17.73	21.66	4.76	0.15	5.69	10.07
+74	5.26	20.68	16.96	21.87	5.00	0.11	4.62	10.99
+53	4.33	22.34	17.46	20.85	5.22	0.11	4.49	11.07
+37	4.47	24.85	19.80	20.40	5.48	0.16	4.85	9.19
+25	2.15	28.24	20.67	19.29	5.38	0.20	5.29	9.14
-25	19.52	11.67	33.16	12.76	7.89	0.26	1.89	10.25

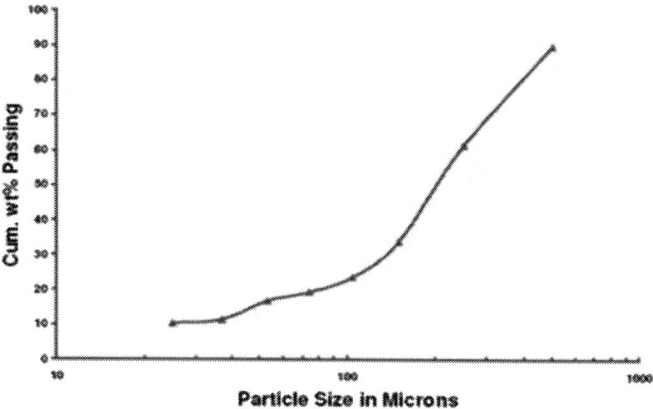

Figure 2: Size analysis of the COB plant tailings.

Mineralogical Characterisation Studies

Mineralogical Characterization studies were carried out for different size fractions using stereomicroscope and reflected light microscope. XRD studies of some of the sieve fractions were also carried out to confirm the mineralogical results. Liberation studies were carried out for the sieve classified samples viz. +500µ; +300µ; + 10µ; +150µ; +100µ and -100µ samples.

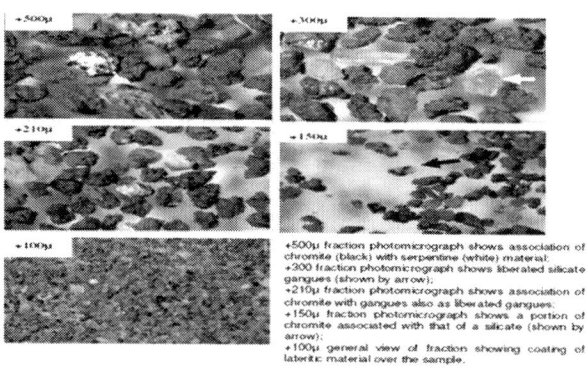

Figure 3: Photomicrographs of the sieve classified samples depicting interlocking of chromite with different types of silicate gangues. All the photomicro-

graphs are taken under stereomicroscope. x16.+500p fraction photomicrograph shows association of chromate (black) with serpentine (white) material: .300 fraction photomicrograph shows kberated silicate gangues (shown by arrow): .210u fraction photomicrograph shows association of chromate with gangues also as iberated gangues: +150p fraction photomicrograph shows a portion of chromate associated with that of a silicate (shown by snow): .100p general view of fraction showing coaling of laterkic material over the sample.

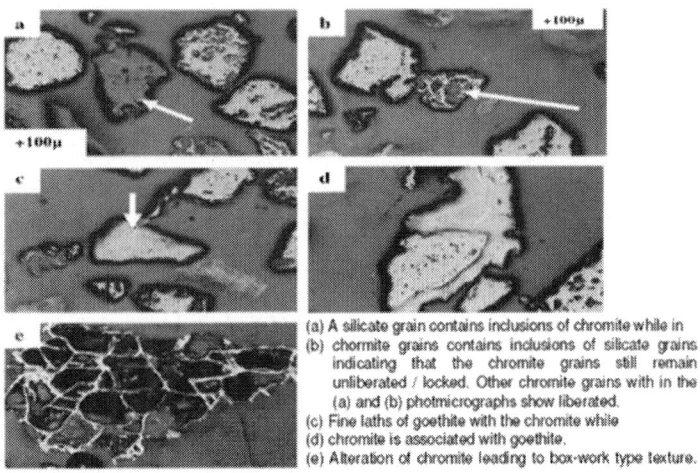

(a) A silicate grain contains inclusions of chromite while in
(b) chormite grains contains inclusions of silicate grains indicating that the chromite grains still remain unliberated / locked. Other chromite grains with in the (a) and (b) photmicrographs show liberated.
(c) Fine laths of goethite with the chromite while
(d) chromite is associated with goethite.
(e) Alteration of chromite leading to box-work type texture.

Figure.4: Photomicrographs of the siev62%ssified samples under reflected light microscope.

Microscopic Studies

Chromite: Chromite occurs as euhedral to subhedral crystal grains, rarely angular and elongated grains are also noticed. Occasionally chromite grains were fractured. Chromite is more enriched in Fe than normal as alternation product from chromite. Rarely the alteration of chromite leads to box-work type texture (Fig.4e). Many times chromite grains are locked either within the iron ore minerals (goethite/hematite) or within silicates or the chromite with inclusions of silicate (Figs.4a and b).

Iron minerals: Goethite and hematite are the two main iron ore minerals. Goethite occurs as massive, colloform bands, botryoids and also as highly friable forming fine matrix. Because of this fineness it gives a lateritic coating on the sample (Fig.3 +100μ). Hematite occurs as irregular masses, streaks, laths, and very intimately and intricately associated with goethite. Hematite contains inclusions of chromite and vice versa (Figs.3c and d).

Silicate: Generally serpentine and quartz form the silicate matrix in the sample. Quartz is coarse grained and liberated at a coarse size (Fig.3, +300 μ). Many a times it is observed that the serpentine (Fig.3; +500μ) and locked with in the chromite grains (Fig.3; +210 μ and +150μ).

Effect of Teeter Water on Cut Size (D_{50})

The relation of cut size (D_{50}) with teeter water flow rate and bed pressure is shown in Figure 5. From the figure it is evident that there is an increase in the bed pressure and teeter water flow rate will increase the separation size or cut size from 35 to 125 microns, which has a large impact on the maximum quantity of particles transport to the overflow. This can be explained that when bed pressure increases, the teeter bed height increases, which is a function of teeter water flow rate and bed pressure, there by pushing the coarse particles to overflow launder will increase i.e the distribution of coarse particle is more in overflow. As a result the cut size increases with increase in the bed pressure and teeter water flow rate.

Effect of Teeter Water on Recovery of Cr_2O_3

The effect of teeter water flow rate on the percentage recovery of Cr_2O_3 at different bed pressure is presented in Figure 6. It can be observed from the Figure 6 that with an increase in the teeter water flow rate from 6lpm to 12lpm there is a decrease in the recovery of Cr_2O_3 in underflow. For example at constant bed pressure (0.06 bar) the recovery to underflow decreases from 93.86% to 86.33%. Similarly at higher bed pressure (0.075 bar) the recovery to underflow is very low i.e 65.92%. This decrease in the recovery of Cr_2O_3 at underflow is attributed due to the increase in the teeter water flow rate through which the fluidized column will become more loosened and the upward water current may force the fine heavies and coarse lights report into the over flow fraction along with the fine light.

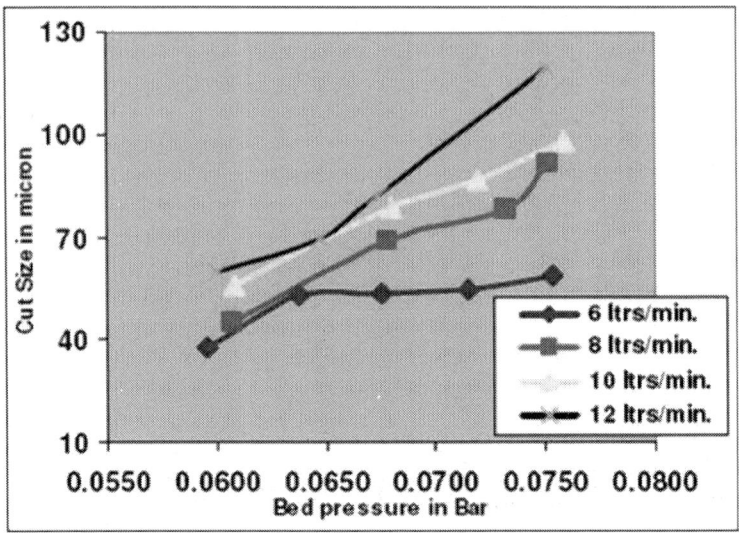

Figure 5: Effect of bed pressure and teeter water on cut size.

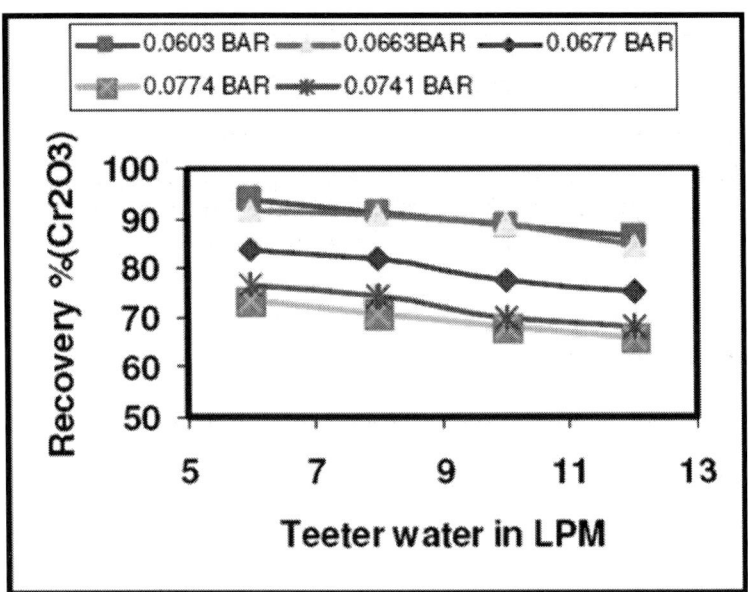

Figure 6: Effect of teeter water flow rate on Recovery Cr_2O_3.

Effect of Bed Pressure

Floatex density separator works based on the principle of hindered settling. Richardson and Zaki proposed the particle slip velocity, which is the relative velocity between the particle and water velocity, is a function of the terminal settling velocity and liquid fraction of the suspension. When the slip velocity of the particle is equal to the interstitial teeter water velocity, the particle will have a zero velocity with respect to stationary viewer and will have an equal chance to report either to overflow or to the underflow stream. Moreover, the interstitial teeter water velocity is related to the voidage (liquid fraction) of the bed, decreases with increase in bed pressure. The pressure setting determines the accumulation of the deposited material inside the unit and hence the bed height inside the FDS. As the bed height increases the average density and viscosity of the suspension goes up increasing the resistance for the particle to settle. Thus the cut size increase which favorably reject the gangue minerals.

Effect of Teeter Bed Pressure on Recovery of Cr_2O_3

The effect of teeter bed pressure on percentage recovery of Cr_2O_3 to underflow is shown in Figure 7. It can be illustrated from the figure that there is a decrease in the recovery of Cr_2O_3 to underflow as teeter bed pressure increases. That shows an increase in the bed height owing to which more loosed fluidized column, the upward water current may force the coarse particles along with the fine chromite particles reporting into the over flow fraction. In other words the coarse lighter and fine heavier particles, having the same slip velocity segregate in the teetered column and push them to overflow fraction. Further this may be elucidated that the bed pressure controls the performance of the classifier.

Classification Performance

Classification is a significant unit operation for separating the mixtures of minerals into two ore more fractions on the basis of the velocity with which the grains fall through a fluid medium. Floatex density separator is a hydraulic classifier which works on the principle of hindered settling and fluidisation. Therefore, the performance is evaluated in terms of imperfection (I) which is defined as

$I = (D_{75} - D_{25}) / 2D_{50}$

Where D_{75} is the particle size where 75% of the mass reported to the underflow, D_{25} is the particle size where 25% of the mass report to the underflow and D_{50} is the cut size where 50% of the mass report either to underflow or overflow. The lower imperfection (I) value indicates better classification efficiency which explains the degree of misplacement of the fine particles to the underflow or the short-circuiting of the feed to overflow without any classification.

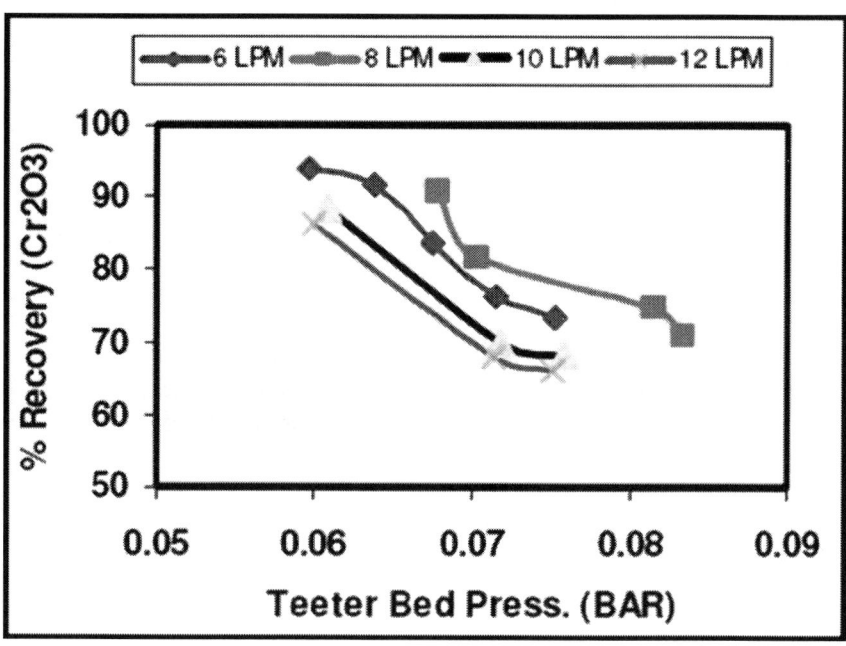

Figure 7: Effect of teeter bed pressure on Recovery Cr_2O_3.

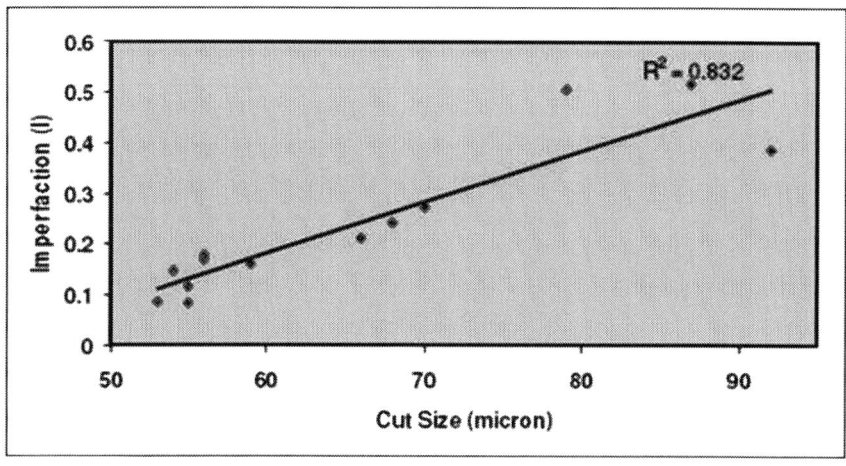

Figure 8: Correlation between cut size and Imperfection.

From the Figure 8 it is clear that, as the cut size (D_{50}) increases the separation efficiency or imperfection (I) increases. This can be explained as the cut size increases the coarse light particles will also be pushed to overflow fraction along with the fine heavy particles. A linear dependence of misplacement of particles i.e imperfection (I) on cut size (D_{50}) is evident from the Figure 8. The correlation coefficient was found to be reasonable at 0.83. As the key operating variables i.e. teeter water flow rate and bed pressure which has the direct impact on cut size. So at higher bed pressure or teeter water the coarse light to the overflow has a predominant in the classification.

CONCLUSIONS

It may be concluded from the results that significant removal of iron bearing mineral such as goethite and silica was possible using FDS. In a single stage operation with FDS, a maximum of 83% recovery of chromite is possible with 22 to 23% Cr_2O_3 content. Thus obtained FDS underflow is suitable for further enrichment using flotation or any other techniques. A low teeter water flow rate with a high bed pressure removes iron bearing mineral like goethite efficiently.

ACKNOWLEDGEMENTS

Authors are thankful to Dr. D. Bhattacharjee, Chief R&D and SS and General Manager, Sukinda for giving an opportunity to work on this project. Special thanks are due to the COB plant personnel especially Mr.R.N.Behra for helping in sample collection and valuable discussion. The support and services provided by staff of R & D division are also duly acknowledged.

REFERENCES

1. Heiskanen, K., 1993, *Particle Classification,* Chapman and Hall, London.
2. De Kok, S.K., 1975, "Fine sizing in milling circuits." *Journal of SAIMM*, October, pp 83-86.
3. Heiskanen, K., 1979, "Two stage classification." *World Mining*, Vol. 32, pp 44-46.
4. Hukki, R.T., and Heiskanen, K., 1981, "Two stage hydrallic classification: A report on industrial application." *110th Annual AIME meeting*, Chicago.
5. Lynch, A.J., 1977, *Mineral crushing and grinding circuits,* Elsevier, New York.
6. Luckie, P., Hogg, R., and Schaller, R., 1980, "A review of two fine particle processing unit operations – Classification and Mixing." *Fine particle processing*, Somasundaran (Ed), Vol.1, pp 167-180.
7. Rogers, R.S.C., Hukki, A.M.., Steiner, G.J., and Arterburn, R.A., 1981, "An evaluation of the use of two vs. one stage hydro cyclones in a pilot scale ball mill." *110th Annual AIME meeting*, Chicago.
8. Guney A., Onal G., and Atmaca T., 2001, *New aspect of chromite gravity tailings reprocessings,* Vol. 14, pp 1527-1530.
9. Rao S.M., Chandrakala K., Kapure G., Nath G., and Rao N.D., 2006, "Recovery of chromite values from chrome ore beneficiation plant tailings." *Tata Search*, pp 59-64.
10. Guney A., Sirkesi A. A., Gurkan V., and Onal G., 1996, "The recovery of chromite fines from the tailings of Uckopru chromium

plant using HIWMS." *Changing scopes in mineral processing*, pp 149-154.
11. Sarkar. B., Das A, and Mehrotra S.P., 2008, "Study of separation features in floatex density separator for cleaning fine coal." *Int. J. Miner. Process.*, Vol. 86, pp. 40–49.
12. Richardson J. F., and Zaki W. N., 1954, *Trans. Inst. Chem. Eng.*, Vol. 32, pp. 35.

Cottonseed Yield and its Quality as Affected by Mineral Fertilizers and Plant Growth Retardants

Zakaria M. Sawan

Cotton Research Institute, Agricultural Research Center, Ministry of Agriculture and Land Reclamation, Giza, Egypt

ABSTRACT

The increase in the population in Egypt makes it imperative to explore promising approaches to increase food supply, including protein and oil, to meet the needs of the Egyptian people. Cotton is the principal crop of Egyptian agriculture, it is grown mainly for its fiber, but cottonseed products are also of economic importance. Cottonseed is presently the main source of edible oil and meal for livestock in Egypt. Economic conditions in modern agriculture demand high crop yields in order to be profitable and consequently meet the high demand for food that comes with population growth. Oil crop production can be improved by development of new high yielding varieties, and the application of appropriate agronomic practices. There is limited information about the most suitable management practice for application of N, P, K, Zn, Ca and PGRs in order to optimize the quantity and quality of oil and protein of cottonseed. In maximizing the quantity and quality of a crop's nutritional value in terms of fatty acids and protein, field experiments were conducted to investigate the effect of nitrogen, phosphorus, potassium, foliar application of zinc and calcium, the use of plant growth retardants (Pix, Cycocel or Alar), on cottonseed, protein, oil yields, and oil properties of Egyptian cotton. From the findings of this study, it seems rational to recommended applied of N, P, K, foliar application of Zn and Ca, the use of PGRs (Pix, Cycocel or Alar), could bring about better impact on cottonseed yield, seed protein content, oil and protein yields, oil refractive index, unsaponifiable matter, and unsaturated fatty acids in comparison with the ordinary cultural practices adopted by Egyptian cotton producers.

INTRODUCTION

Plant nutrition, using a balanced fertilization program with both macro and micro-nutrients is becoming necessary in the production of high yield with high quality products especially with the large variation in soil fertility and the crop's need for macro and micro-nutrients. The breeding and production of cotton have traditionally been guided by consideration of fiber quality and yield. However, cottonseed characteristics except for viability and vigor have generally been ignored. Cottonseed oil is an important source of fat. Also, cottonseed meal is classed as a protein supplement in the feed trade and is almost as important as soybean meal [1].

Nitrogen

In cotton culture, nitrogen (N), has a necessity role in production inputs, which controls growth and prevents abscission of squares and bolls, essential for photosynthetic activity [2], and stimulates the mobilization and accumulation of metabolites in newly developed bolls, thus increasing their number and weight. Additionally, with a dynamic crop like cotton, excess N serves to delay maturity, promote vegetative tendencies, and usually results in lower yields [3]. Therefore, errors made in N management that can impact the crop can be through either deficiencies or excesses. Ansari and Mahey [4] evaluate the effects of N level (0, 40, 80, 120 and 160 kg·ha^{-1}) on the yield and found that seed yield increased with increasing N level up to 80 kg·ha^{-1}. Nitrogen is an essential nutrient for the synthesis of fat, which requires both N and carbon skeletons during the course of seed development [5]. On the other hand, nitrogen plays the most important role in building the protein structure [6]. Another beneficial change in fatty acid composition due to N nutrition would be an increase in the linoleic and oleic acid contents, and an increase in the percentage of unsaturated fatty acids and a decrease in saturated fatty acids in the seed oil [7].

Phosphorus

Phosphorus (P) is the second most limiting nutrient in cotton production after nitrogen. Its deficiency tends to limit the growth of cotton plants, especially when plants are deprived of P at early stages than later stages of growth. Further, P is an essential nutrient and an integral component of several important compounds in plant cells, including the sugar-phosphates involved in respiration, photosynthesis and the phospholipids of plant membranes, the nucleotides used in plant energy metabolism and in molecules of DNA and RNA [8]. Phosphorus deficiency reduces the rate of leaf expansion and photosynthetic rate per unit leaf area [9]. Sasthri et al. [10] found that application of 2% diammonium phosphate to cotton plants increased seed yield. Improvements in cotton yield resulting from P application were reported by Stewart et al. [11]; Singh et al. [12]; Ibrahim et al. [13]; Gebaly and El-Gabiery [14].

Potassium

The physiological role of potassium (K) during fruit formation and maturation periods is mainly expressed in carbohydrate metabolism and translocation of metabolites from leaves and other vegetative organs to developing bolls. Potassium increases the photosynthetic rates of crop leaves, CO_2 assimilation and facilitating carbon movement [15]. The high concentration of K^+ is thought to be essential for normal protein synthesis. Potassium deficiency during the reproductive period can limit the accumulation of crop biomass [16], markedly changes the structure of fruit-bearing organs, and decreases yield and quality. Improvements in cotton yield and quality resulting from K input have been reported by the following authors: Gormus [17] applying K rates of 66.4, 132.8 and 199.2 kg·ha^{-1} K; Aneela et al. [18] increase K levels, the effect being highest at 166 kg·K·ha^{-1}; Pervez et al. [19] using K rates of 62.5, 125 and 250 kg·ha^{-1}; Pettigrew et al. [20] with a K fertilizer rate of 112 kg·ha^{-1}; Sharma and Sundar [21] with a foliar application of K at 4.15 kg·ha^{-1}.

Zinc

Zinc (Zn) is critical for several key enzymes in the plant. Zinc binds tightly to Zn-containing essential metabolites in vegetative tissues, e.g., Zn-activated enzymes such as carbonic anhydrase [22]. Further, Zn is required in the biosynthesis of tryptophan, a precursor of the auxin—indole-3-acetic acid (IAA), which is the major hormone inhibiting abscission of squares and bolls. Zinc deficiency symptoms include: small leaves, shortened internodes, a stunted appearance, reduced boll set, and small bolls size [23]. Zinc deficiency occurs in cotton on high-pH soils, and where high rates of P are applied [23]. Rathinavel et al. [24] found that application of $ZnSO_4$, to the soil at 50 kg·ha^{-1} increased 100-seed weight. Li et al. [25] found that when cotton was sprayed with 0.2% zinc sulfate at the seedling stage, the boll number plant^{-1} increased by 17.3% and the cotton yield increased by 18.5% compared with the untreated control.

Calcium

Calcium (Ca) is essential in cell nucleus matrix. It activates enzymes, particularly those that are membranebound [26]. It is thought that Ca is important in the formation of cell membranes and lipid structures. Ma and Sun [27], suggested that Ca might be involved in light signal transduction chain for phototropism. Calcium deficiency as one of the causes of abscission and suggested this plus the role of Ca in the middle lamella (Ca pectates) as the possible reason. A likely reason was that Ca deficiency affected translocation of carbohydrates, causing accumulation in the leaves and a decline in stems and roots. It seems probable that young bolls abscised because of starvation. Thus, Ca may inhibit abscission because it is a component of the middle lamella, because it promotes translocation of sugars and auxin, and because it helps prevent senescence. Ochiai [28] notes that Ca^{2+} can bridge phosphate and carboxylate groups of phospholipids and proteins; that it increases hydrophobicity of membranes; that it generally increases membrane stability and reduces water permeability.

Plant Growth Regulators Retardants

An objective for using plant growth retardants (PGRs) (mepiquat chloride, "Pix", chloromequat chloride, "Cycocel", and daminozide, "Alar") in cotton is to balance vegetative and reproductive growth as well as to improve yield and its quality [29]. Visual growth-regulating activity of Pix, Cycocel or Alar is similar, being expressed as reduced plant height and width, shortened stem and branch internodes and leaf petioles, influence leaf chlorophyll concentration, structure and CO_2 assimilation, and thicker leaves. This indicates that bolls on treated cotton plants have a larger photo synthetically sink for carbohydrates and other metabolites than those on untreated plants. More specific response from using PGRs include alteration of carbon partitioning, greater root/ shoot ratios, enhanced photosynthesis, altered nutrient uptake, and altered crop canopy. In this connection, Wang et al. [30] stated that application of the plant growth retardant Pix to the cotton plants at squaring decreased the partitioning of assimilates to the main stem, the branches and their growing points, and increased partitioning to the reproductive organs and roots. Also, they indicated that, from bloom to boll-setting, Pix application was very effective in restricting the vegetative growth of the cotton canopy and in promoting the partitioning of assimilates into reproductive organs. Kumar et al. [31] evaluated the effects of Chamatkar, 5% Pix, 500, 750 and 1000 ppm, on cotton. These treatments increased the values for photosynthetic rate, transpiration rate, total chlorophyll content, and nitrate reductase activity, number of bolls $plant^{-1}$, boll weight and yield.

Most previous has focused on studying the effect of nitrogen, phosphorus, potassium, foliar application of zinc and calcium, the use of PGRs on cotton yield and fiber quality [32,33]. However, there is limited information about the most suitable management practice for application of N, P, K, Zn, Ca and PGRs in order to optimize the quantity and quality of oil and protein of cottonseed [34]. Due to the economic importance of cottonseed (presently the main source of edible oil and meal for livestock) in Egypt, this study was designed to identify the best combination of these production treatments in order to improve cottonseed, protein and oil yields and oil properties of Egyptian cotton (G. barbadense L.) [1,35-37].

METHODS AND MEASUREMENTS

Field experiments were conducted at the Agricultural Research Center, Ministry of Agriculture in Giza (30°N, 31°: 28'E and 19 m altitude), Egypt using the cotton cultivars "Giza 75" and "Giza 86" (Gossypium barbadense L.) in the two seasons I and II. Seeds were planted on March, and seed cotton was harvest on October [38]. The soil type was a clay loam. Average textural and chemical properties of soil are reported in Table 1 [38]. Range and mean values of the climatic factors recorded during the growing seasons are presented in Table 2 [38]. No rainfall occurred during the two growing seasons. The experiments were arranged as a randomized complete block design. The plot size was 1.95 × 4 m, including three ridges (beds). Hills were spaced 25 cm apart on one side of the ridge, and seedlings were thinned to two plants hill^{-1} 6 weeks after planting, providing plant density of 123,000 plants·ha^{-1}. Total irrigation amount during the growing season (surface irrigation) was about 6000 m^3·ha^{-1}. The first irrigation was applied 3 weeks after sowing, and the second one was 3 weeks later. Thereafter, the plots were irrigated every 2 weeks until the end of the season, thus providing a total of nine irrigations [38].

Experiments

Effect of N, Zn and PGR's on Cottonseed, Protein, Oil Yields, and Oil Properties

Materials: A field experiment was conducted, using the cotton cultivar Giza 75. Each experiment included 16 treatments the following combinations: 1) Two N-rtes (107 and 161 of N·ha^{-1}) were applied as ammonium nitrate (33.5% N) in two equal amounts 6 and 8 weeks after sowing; each application (in the form of pinches beside each hill) was followed immediately by irrigation. 2) Three PGR's, 1, 1-dimethylpiperidinium chloride (mepiquat chloride, or Pix), 2-chloroethyltrimethylammonium chloride (chloromequat chloride, or Cycocel), and succinic acid 2, 2-dimethylhydrazide (daminozide, or Alar) were used. Each was foliar-sprayed once at 288 g active ingredient ha^{-1}, 75 days after planting (during square initiation and boll setting stage) at solution volume of 960 l·ha^{-1}. Water was used as the control treatment. 3) Two chelated Zn rates (0.0 and 48 g of Zn ha^{-1}) were foliar-sprayed twice, 80 and 95 days after planting at solution volume of 960 l·ha^{-1} [35].

Effect of P, Zn and Ca on Cottonseed, Protein and Oil Yields and Oil Properties

Materials: A field experiment was conducted on the cotton cultivar Giza 75. Each experiment included 16 treatments, using combinations: 1) Two P rates, 44 (farmer's dose) and 74 kg of P$_2$O$_5$ ha^{-1} were applied (as a concentrated band close to the seed ridge) as calcium super-phosphate (15% P$_2$O$_5$) before the first irrigation, i.e. 3 weeks after planting (during seedling stage). 2) Two Zn rates at 0.0 and 40 ppm, as chelated form [ethylenediaminetetraacetic acid (EDTA)] each was foliar sprayed twice, 75 and 90 days after planting (during square initiation and boll setting stage) at solution volume of 960 l·ha^{-1}. 3) Four chelated Ca rates at 0.0, 20, 40 and 60 ppm were each foliar sprayed twice, 80 and 95 days after planting, at solution volume of 960 l·ha^{-1} [36].

Cottonseed, Protein, Oil Yields, and Oil Properties as Influenced by K, P and Zn

Materials: A field experiment was conducted on the cotton cultivar "Giza 86". Each experiment included 16 treatment combinations of the following: 1) Two K rates (0.0 and 47.4 kg of K ha^{-1}) were applied as K sulfate (K_2SO_4, "48% K_2O), eight weeks after sowing (as a concentrated band close to the seed ridge) and the application was followed immediately by irrigation. 2) Two Zn rates (0.0 or 57.6 g of Zn ha^{-1}) were applied as chelated form and each was foliar sprayed two times (70 and 85 days after planting, during square initiation and boll setting stage). 3) Four phosphorus rates (0.0, 576, 1152 and 1728 g of P ha^{-1}) were applied as calcium super phosphate (15% P_2O_5) and each was foliar sprayed two times (80 and 95 days after planting).

Table 1: Physical and chemical properties of the soil used in I and II seasons

Season	I	II
Sail texture		
Clay (%)	43.0	46.5
Silt (%)	28.4	26.4
Fine sand (%)	19.3	20.7
Coarse sand (%)	4.3	1.7
Soil texture	Clay	Loom
Chemical analysis		
Organic matter (%)	1.8	1.9
Calcium carbonate (%)	3.0	2.7
Total soluble salts (%)	0.13	0.13
pH(1:2.5)	8.1	8.1
Total nitrogen (%)a	0.12	0.12
Available nitrogen (mg-kg-1 soil)b	50.0	57.5
(1% K2SO4, extract)		
Available phosphorus (mg- kg-1 soil)	15.7	14.2

(NaHCO3 0.5 N, extract) Available potassium (mg kg-1 soil)	370.0	385.0
(NH4OAC 1 N, extract) Total Sulphur (mg.kg-1 soil)	21.3	21.2
Calcium (meq/100 g) (with Virsen, extract)	0.2	0.2

[a]Total nitrogen, i.e. organic N + inorganic N. [b]Available nitrogen, i.e. NH_4^+ & NO_3^-. The Physical analysis (soil fraction) added to the organic matter, calcium carbonate and total soluble salts to a sum of about 100% [38].

Table 2: Range and mean values of the weather variables recorded during the growing seasons (April-October)

Weather variables	Season I		Season II		Overall date (Two seasons)	
	Rang	Mean	Range	Mean	Range	Mean
Max Temp [t]	20.8 - 44.0	32.6	24.6 - 43.4	32.7	20.8 - 44.0	32.6
Min Temp [°C]	10.4 - 24.5	19.4	12.0 - 24.3	19.3	10.4 - 24.5	19.3
Max-Min Temp [t]	4.7 - 23.6	132	8.5 - 26.8	13.4	4.7 - 26.8	13.3
Sunshine [h. d-1]	0.3 - 12.9	11.1	1.9 - 13.1	11.2	0.3 - 13.1	11.1
Max Hum [%]	48 - 96	79.5	46 - 94	74.7	46 - 96	77.2
Min Hum [%]	6-48	30.1	8-50	310	6-50	31.5
Wind speed [m.s-1]	0.9 - 11.1	5.2	1.3 - 11.1	5.0	0.9 - 11.1	5.1

The Zn and P were both applied to the leaves with uniform coverage at a solution volume of 960 l·ha⁻¹, using a knapsack sprayer [1].

Effects N, K and PGR on Oil Content and Quality of Cotton Seed

Materials: A field experiment was conducted, using the cotton cultivar "Giza 86". The experiment included 16 treatments: 1) soil application of N (95.2 "the ordinary", and 142.8 kg of N ha^{-1} as ammonium nitrate), 2) foliar application of K (0, 319, 638 and 957 g·K·ha^{-1} as potassium sulfate) and 3) foliar spray of the PGR (1,1-dimethylpiperidinium chloride (mepiquat chloride "MC" or "Pix") 75 days after planting at 0 or 48 g a.i. ha^{-1}, and 90 days after planting at 0 or 24 g a.i. ha^{-1}). The solution volume applied was also 960 L·ha^{-1}. Nitrogen fertilizer (NH_4NO_3, "3.5% N") was applied half at 6 and the rest at 8 weeks after planting. The fertilizer was placed beside each hill in the form of pinches and followed immediately by irrigation. Potassium (K_2SO_4, "40% K") was applied as foliar spray during square initiation and boll development stage, 70 and 95 days after planting, respectively. The solution volume applied was 960 l·ha^{-1}. The K and MC were applied to the leaves uniformly using a knapsack sprayer [37].

Measurements

At harvest the seed cotton yield plot^{-1} (handpicking) was determined. Following ginning, the cotton seed yield in kg·ha^{-1} as well as 100-seed weight in g was determined. A composite seed sample was collected from each treatment for chemical analyses. The following chemical analyses were conducted: 1) seed crude protein content according to AOAC standards [39]; 2) seed oil content in which oil was extracted three times with chloroform/methanol (2:1, vol/vol) mixture according to the method outlined by Kates [40]; 3) oil quality traits, i.e., refractive index, acid value, saponification value, unsaponifiable matter, and iodine value were determined according to methods described by AOCS [41]; and 4) identification and determination of oil fatty acids by gasliquid chromatography. The lipid materials were saponified, unsaponifiable matter was removed, and the fatty acids were separated after acidification of the saponifiable materials. The free fatty acids were methylated with diazomethane [42]. The fatty acid methyl esters were analyzed by a Hewlett Packard model 5890 gas chromatograph (Palo Alto, CA) equipped with dual flame-ionization detectors. The separation procedures were similar to those reported by Ashoub et al. [43].

Statistical Analysis

Data obtained for the cottonseed yield and seed weight were statistically analyzed as a factorial experiment in a RCBD following the procedure outlined by Snedecor and Cochran [44] and the least significant difference (LSD) was used to determine the significance of differences between treatment means. As for the chemical properties considered in the study, the t-test computed in accordance with standard deviation was utilized to verify the significance between treatments means [1].

ANALYZED DATA FOR MEASUREMENTS

Experiments

Effect of N, Zn and PGR's on Cottonseed, Protein, Oil Yields, and Oil Properties

Seed yield ha^{-1}, was significantly (P ≤ 0.05) increased (8.96%) by raising N rate (Table 3) [35]. Abdel-Malak et al. [45] stated that cotton yield was higher when N was applied at a rate of 190 kg·ha^{-1} than at the rate of 143 kg·ha^{-1}. Palomo Gil and Chávez González [32] applied N at a rate ranging from 40 to 200 kg·ha^{-1} to cotton plants and found highest yield was associated with high rates of applied N. Similar results were obtained by Sarwar Cheema et al. [33] Saleem et al. [46] when N was applied at 120 kg·ha^{-1}; Hamed et al. [47] when N was applied up to 178 kg·ha^{-1}. Nitrogen is an important nutrient which control growth and prevents abscission of squares and bolls, essential for photosynthetic activity [2] and stimulate the mobilization and accumulation of metabolites in newly developed bolls and thus their number and weight are increased. All tested PGR (Pix, Cycocel and Alar) significantly increased seed yield ha^{-1} (7.79% - 12.08%), compared with the untreated control. The most effective was Pix (12.08%), followed by Cycocel (10.57%) [35]. These results may be attributed to the promoting effect of these substances on numerous physiological processes, leading to improvement of all yield components. Pix applications increases CO_2 uptake and fixation in cotton plant leaves. In cotton stems, the xylem was expanded with Pix treatment, perhaps increasing the transport ability and accounting for heavier bolls. Alar and Pix also have been associated with increased photosynthesis [48,49] through increased total chlorophyll concentration of plant leaves, increased photosynthesis greatly increased flowering, boll retention and yield. Abdel-Al [50] indicated that cotton yield significantly increased with Pix treatment at a rate 11.90 ml (formulation) ha^{-1} at the beginning of flowering, and Gebaly and El-Gabiery [14] found that cotton yield significantly increased with Pix application at 1, 2 and 3 cm^3 L (formulation) at pinhead square, start of flowering and peak of flowering stage. Pípolo et al. [51] found that spraying cotton plants at an age of 70 d after emergence with Cycocel at rates ranging from 25 to 100 $g \cdot ha^{-1}$ resulted in yield increases. Sawan et al. [52] stated that application of Cycocel and Alar, at rates ranging from 250 to 700 ppm (105 days after planting) increased cotton seed yield ha^{-1}. Similar results were obtained by Sarwar Cheema et al. [33]. Application of Zn significantly increased seed yield ha^{-1} (8.44%), as compared with untreated plants [35]. Zeng [53] stated application of Zn to cotton plants on calcareous soil increased yield by 7.8% - 25.7%. Similar results were obtained by Ibrahim et al. [13]. Zinc is required in the synthesis of tryptophan, which is a precursor of IAA synthesis which is the hormone that inhibits abscission of squares and bolls. Also, this nutrient has favorable effect on the photosynthetic activity of leaves and plant metabolism [25], which might account for higher accumulation of metabolites in reproductive organs (bolls).

Seed index significantly increased with increasing N rate (Table 3) [35]. This may be partially due to enhanced photosynthetic activity [2]. Similar findings were obtained by Palomo Gil and Chávez González [32];

Table 3: Effect of N rate and foliar application of plant growth retardants and Zn on cottonseed yield, seed index, seed oil, seed protein, oil and protein yields

Treatments		Cottonseed yield (kg ha-1)a	Seed index (g)a	Seed Oil (%)b	Oil yield (kg ha-1)b	Seed protein (%)b	Protein yield (kg. ha-1)b
N-rate (kg.ha-1)							
Control	107	1907.7	10.29	19.92	380.1	21.96	418.8
	161	2078.7	10.44	19.87	413.0	22.51	468.4
L.S.D. 0.05c		67.8	0.06		-		
S.E.d		-	-	0.02	16.4	0.27	24.8
Plant growth retardants (ppm)							
Control	0	1852.2	10.24	19.86	368.0	22.08	409.1
Pix	300	2076.0	10.44	19.88	413.2	22.35	465.0
Cycocel	300	2048.0	10.41	19.94	407.8	22.24	455.8
Alar	300	1996.5	10.36	19.90	397.3	22.26	444.5
L.S.D. 0.05c		96.0	0.08	-	-	-	-
S.E.d		-	-	0.01	10.1	0.05	12.2
Zn rate (ppm)							
Control	0	1912.4	10.30	19.82	379.2	22.10	422.8
	50	2073.9	10.42	19.97	413.9	22.37	464.4
L.S.D. 0.05c		67.8	0.06	-	-	-	-
S.E.d		-	-	0.07	17.3	0.13	20.8

aCombined statistical analysis from the two seasons. bMean data from a four replicate composite for the two seasons. dL.S.D. = Least significant differences, cS.E. = standard error. [35].

Hamed et al. [47]. Application of all PGR significantly increased seed index as compared to untreated control; Pix gave the highest seed index, followed by Cycocel [35]. These agree with previous works of Sawan et al. [52], by applying Cycocel and Alar; Carvalho et al. [54] by applying Pix and Cycocel; Abdel-Al [50], by applying Pix. Zinc significantly increased seed index compared with the untreated control [35]. In this connection Ibrahim et al. [13] noted that seed weight increased due to the application of Zn.

Seed oil content was unchanged with increased as N-rate. Oil yield ha^{-1} significantly (32.9 kg·ha^{-1}), which is attributed to the increase in seed yield (Table 3) [35]. Pandrangi et al. [55] applied N at a rate of 25 or 50 kg·ha^{-1} to cotton plants and found that the percentage of seed oil content decreased but oil yield increased with increasing N rate. Application of all growth retardants resulted in an insignificant increase in seed oil content above the control and also significantly increased the oil yield ha^{-1} over the control (29.3 - 45.2 kg oil ha^{-1}), with the clearest effect from Pix (45.2 kg·ha^{-1}), followed by Cycocel (39.8 kg·ha^{-1}) [35]. Sawan et al. [56] indicated that a slight increase in cottonseed oil content was detected with Pix application at rate ranging 10 - 100 ppm. Pix was sprayed once (90 D) or twice (90 and 110 days after planting). Oil yield also increased due to Pix application compared with the control. Similar results were obtained by Gebaly and El-Gabiery [14]. Sawan et al. [52] observed that application of Cycocel and Alar (250 - 750 ppm, 105 days after planting) increased oil yield ha^{-1}. Application of Zn resulted in an insignificant increase in seed oil content over that of the control. The seed oil yield was also increased (34.7 kg oil ha^{-1}) compared with the untreated control [35]. These results could be attributed to the increase of total photoassimilates (e.g. lipids) and the translocated assimilates to the sink as a result of applying Zn nutrient. Sawan et al. [57] found that oil yield increased by the application of Zn to cotton plants at a rate of 12 g Zn ha^{-1}. Zinc was sprayed three times, i.e., 70, 85, and 100 d after sowing. Prabhuraj et al. [58] found that applying Zn at 5 ppm rate increased seed and oil yields of sunflower. Similar results were obtained by Ibrahim et al. [13] on cotton; Bybordi and Mamedov [59] on canola.

High N rate significantly increased the seed protein content and yield ha^{-1} (49.6 kg protein ha^{-1}) (Table 3) [35]. According to Sugiyama et al. [60], soluble proteins are increased with better N supply and favorable growth condition. These results suggest that the high N-rate increases the amino acids synthesis in the leaves, and this stimulates the accumulation of protein in the seed rather than oil content. Patil et al. [5] found that N application (50 kg·N·ha^{-1}) increased the seed protein content. Seed protein content and yield ha^{-1} were increased insignificantly in plants in plants treated with the three growth retardants (35.4 - 55.9 kg protein ha^{-1}) compared with the untreated control. Highest protein content was produced by Pix application, followed by Alar, while the highest seed protein yield was obtained with Pix

(55.9 kg·ha^{-1}), and followed by Cycocel (46.7 kg·ha^{-1}) [35]. Hedin et al. [61] found that Cycocel increased protein content by 17% - 50% in leaves and squares harvested 4 wk after the first application. Kar et al. [62] in safflower showed that Cycocel and Alar maintained the level of chlorophyll, protein, and RNA contents. Also, the increase in seed protein content may be caused by the role of Pix in protein synthesis, encouraging the conversion of amino acids into protein [63]. Sawan et al. [56]; Gebaly and El-Gabiery [14] stated that cottonseed protein content and yield ha^{-1} increased due to the application of Pix. Kler et al. [64] found that when cotton was sprayed using Cycocel rates of 40, 60, or 80 ppm at the age 63 days after planting, seed protein content increased. Sawan et al. [52] stated that application of Cycocel or Alar increased seed protein content and protein yield ha^{-1}. Application of zinc increased insignificantly the seed protein content and significantly increased protein yield ha^{-1} (41.6 kg protein ha^{-1}) over the untreated control [35]. In these circumstances Ibrahim et al. [13] found that application of Zn to cotton plants increased seed protein content and protein yield ha^{-1}.

The seed oil refractive index and unsaponifiable matter tended to increase insignificantly, while the oil acid value and saponification value tended to decrease by raising N-rate (Table 4) [35]. The increase in unsaponifiable matter is beneficial as it increases the oil stability. Sawan et al. [57] applied N to cotton plants at rates of 108 and 216 kg·ha^{-1} and found that oil unsaponifiable matter tended to increase, while saponification value tended to decrease by raising N-rate. Application of all PGR significantly increased the oil refractive index. However, unsaponifiable matter was insignificantly increased, whereas acid value and saponification value tended to decrease insignificantly as compared with the untreated control [35]. Applied Cycocel gave the highest refractive index and the lowest acid value, while Pix gave the highest unsaponifiable matter. Also, applied Alar gave the lowest saponification value. Sawan et al. [32] stated that application of Cycocel and Alar to cotton plants increased oil refractive index and unsaponifiable matter and decreased oil acid value and saponification value. Osman and Abu-Lila [65] found a negligible variation in refractive index of flax oil when the plants were treated with Cycocel at the application rates of 25 - 100 ppm twice; the first one 20 d after sowing and the second spray 2 months later. The oil refractive index and unsaponifiable matter tended to increase insignificantly, while acid value and saponification value decreased insignificantly by applied zinc compared with control [35]. Sawan et al. [57] found that application of Zn to cotton plants exhibited negligible effect upon oil-quality characters, i.e., refractive index, oil acid value, unsaponifiable matter, and saponification value.

The oil saturated fatty acids (capric, myristic, palmitic and stearic) decreased insignificantly, while lauric acid increased insignificantly in response to raising the N-rate (Table 5) [35]. Palmitic acid was the dominant saturated fatty acid.

Table 4: Effect of N rate and foliar application of plant growth retardants and Zn on seed oil properties[a]

Treatments		Refractive index	Acid value	Saponification value	Unsaponifiable matter (%)
N-rate(kg. ha-1)					
Control	107	1.4733	0.1336	1933	0.3700
	161	1.4734	0.1310	191.6	03738
S.E.b		0.0001	0.0013	1.0	0.0019
Plant growth retardant (ppm)					
Control	0	1.4729	0.1338	193.4	0.3675
Pix	300	1.4734	0.1327	192.9	0.3750
Cycocel	300	1.4738	0.1312	193.1	03725
Alar	300	1.4735	0.1317	191.2	0.3725
S.E.b		0.0002	0.0005	0.5	0.0015
Zn rate (ppm)					
Control	0	L4732	0.1325	193.8	0.3688
	50	L4735	0.1322	191.6	0.3750
S.Eb		0.0001	0.0001	1.1	0.0031

[a]Mean data from a four replicate composite for the two seasons. [b]S.E. = standard error [35]

Low content of saturated fatty acids is desirable for edible uses. Application of the three PGR's resulted in a decrease in the total saturated fatty acids compared with the untreated control. The decrease was significant with the Cycocel and Alar treatments. Cycocel gave the lowest total saturated fatty acids in oil contents, followed by Alar and also tended to increase insignificantly the saturated fatty acid capric acid compared with the untreated control. Applied Pix gave the highest capric and the lowest stearic acid content, while applied Cycocel gave the lowest lauric acid content. Alar application tended to give the lowest myristic and palmitic acids contents compared with control. Application of Zn resulted in a significant decrease in the total saturated fatty acids (capric, palmitic and stearic) while it resulted in an increase in the lauric and myristic saturated fatty acids, compared with untreated plants [35].

The total unsaturated fatty acids (oleic and linoleic) and the ratio between total unsaturated fatty acids and total saturated fatty acids (TU/TS) increased insignificantly (3.53 and 15.93%, respectively) by raising N-rate (Table 6) [35]. Linoleic acid was the most abundant of the unsaturated fatty acids. Kheir et al. [7] found that the higher N-rate increased the percentage of unsaturated fatty acids and decreased saturated fatty acids of flax oil. All PGR's increased the total unsaturated fatty acids and TU/TS ratio, compared with the control. The increase was significant by the application of Cycocel and Alar. Applied Cycocel gave the highest linoleic acid content, total unsaturated fatty acids (10.64%), and TU/TS ratio (51.0%), and followed by Alar (10.02 and 47.01%, respectively) [35]. The increase in TU/TS as a result of the application of the three PGR may be attributed to their encouraging effects on enzymes that catalyzed the biosynthesis of the unsaturated fatty acids. Spraying plants with Zn significantly increased the total unsaturated fatty acids (7.4%) and TU/TS ratio (35.04%), compared with untreated control [35]. Sawan et al. [56] reported that applying Pix to cotton plants caused a general decrease in oil saturated fatty acids, associated with an increase in oil unsaturated fatty acids. Sawan et al. [52] stated that application of Cycocel and Alar to cotton increased oil unsaturated fatty acids. Osman and Abu-Lila [65] when applied Cycocel at rates of 25 - 100 ppm to flax plants found that generally the higher concentrations (50 and 100 ppm) caused in the total oil saturated fatty acids, while they increased the unsaturated fatty acids.

Effect of P, Zn and Ca on Cottonseed, Protein and Oil Yields and Oil Properties

Seed yield ha^{-1} was significantly increased (11.24%) when phosphorus was applied at the highest rate (Table 7) [36]. Phosphorus as a constituent of cell nuclei is essential for cell division and development of meristematic tissue, and hence it should have a stimulating effect on the plants, increasing the number of flowers and bolls per plant. Further, P has a well known impact in photosynthesis as well as synthesis of nucleic acids, proteins, lipids and other essential compounds [66], all of which are major factors affecting boll weight and consequently cottonseed. These results are confirmed by those of Abdel-Malak et al. [45]; Ibrahim et al. [13]; Saleem et al. [67]; Gebaly and El-Gabiery [60]. Application of Zn significantly increased cottonseed yield ha^{-1} (8.61%), as compared with the untreated control [36]. This may be due to its favorable effect on photosynthetic activity, which improves mobilization of photosynthates and directly influences of boll weight [68].

Table 5: Effect of N rate and foliar application of plant growth retardants and Zn on the relative percentage of saturated fatty acids[a]

Treatments			Relative % of saturated fatty acids					
			Capric	Lauric	Myristic	Palmitic	Stearic	Total
N-rate (kg• ha-l)								
Control		107	0.5887	0A375	03700	2032	2.767	25.283
		161	03212	0.8212	0.6812	18.67	2.152	22.646
S.Eb			0.1337	0.1918	0.0444	L02	0307	1319
Plant growth retardants (ppm)								
Control		0	0.3350	12325	1 A050	23.06	2.427	28.459
Pix		300	0.7500	0.7125	0.9225	20.88	1.982	25.247
Cycocel		300	0.3600	0.2600	0.3200	17.59	2.327	20.857
Alar		300	0.3750	03125	02550	1725	1102	21.294
S.Eb			0.0986	0.2250	0.2717	1.38	0.234	1.794

Ca rate (ppm)							
Control	0	0.6325	0.5825	0.5825	22.41	2472	26.679
	50	0.2775	0.6762	0.8687	16.98	2.447	21.249
S.Eb		0.1775	0.0468	0.1431	231	0.012	2.715

[a]Mean data from a four replicate composite for the two seasons. [b]S.E. = standard error [35].

Table 6: Effect of N rate and foliar application of plant growth retardants and Zn on the relative percentage of unsaturated fatty acids[a]

Treatments		Relative % of unsaturated fatty acids			TU/TS[b] ratio
		Oleic	Linoleic	Total	
N rate (kg.ha-1)					
Control	107	21.67	53.04	74.71	2.95
	161	22.57	54.78	77.35	3.42
S.E.c		0.45	0.87	1.32	023
Plant growth retardants (ppm)					
Control	0	20.67	50.86	71.53	2.51
Pix	300	21.20	53.55	74.75	2.96
Cycocel	300	23.07	56.07	79.14	3.79
Alar	300	23.54	55.16	78.70	3.69
S.E.c		0.69	1.14	1.79	0.30
Zn rate (ppm)					
Control	0	21.46	51.86	73.32	2.74
	50	22.79	55.96	78.75	3.70
5.0		0.66	2.05	2.71	0.48

[a]Mean data from a four replicate composite for the two seasons. [b]TU/TS ratio = (total unsaturated fatty acids)/(total saturated fatty acids). [c]S.E. = standard error [35].

Table 7: Effect of P rate and foliar application of Zn and Ca on cottonseed yield, seed index, seed oil, seed protein, oil and protein yields

Treatments		Cottonseed yield (kg• ha-1)a	Seed index (g)a	Seed Oil (%)b	Oil Yield (kg. ha-1) b	Seed protein (%)b	Protein yield (kg ha-1)b
P2O5 rate (kg• ha-1)							
Control	44	1837.1	10.19	19.67	361.6	22.35	410.6
	74	2043.5	10.40	19.86	406.0	22.38	457.5
L.S.D. 0.05c		41.2	0.05	-			
S.E.d				0.09	22.2	0.01	23.4
Zn rate (ppm)							
Control	0	18602	10.24	19.59	364.5	22.22	413.4
	40	2020.4	10.36	19.94	403.0	22.51	454.7
L.S.D. 0.05c		41.2	0.05				-
S.E.d			-	0.17	19.2	0.14	20.6
Ca rate (ppm)							
Control	0	1807.1	10.16	19.74	356.8	22.43	405.3
	20	1934.6	10.31	19.76	382.7	22.36	432.9
	40	1992.7	10.34	19.75	394.2	22.34	445.5
	60	2026.8	10.37	19.82	401.3	22.34	452.4
L.S.D. 0.05c		58.2	0.07				-
S.E.d				0.01	9.7	0.02	10.3

[a]Combined statistical analysis from the two seasons. [b]Mean data from a four replicate composite for the two seasons. [c]L.S.D. = Least significant differences. [d]S.E. = standard error [36].

Also, Zn enhances the activity of tryptophan synthesis, which is involved in the synthesis of the growth control compound IAA, the major hormone that inhibits abscission of squares and bolls. The application of Zn increased the number of retained bolls plant^{-1}. Similar results were obtained by Zeng [53]; Ibrahim et al. [13] on cotton; Bybordi and Mamedov [59] on canola. Calcium application also significantly increased seed yield (7.06% - 12.16%), as yields resulting from the three concentrations applied exceeded the control.

In general, it can be stated that the highest Ca concentration (60 ppm) was more effective than the other two concentrations (20 or 40 ppm) [36]. The role of Ca in increasing seed yield can possibly be ascribed to its involvement in the process of photosynthesis and the translocation of carbohydrates to young bolls. Calcium deficiency depressed the rate of photosynthesis (rate of CO_2 fixation). Guinn [66] stated that Ca deficiency would cause carbohydrates to accumulate in leaves and not in young bolls. The results obtained agree with those reported by Shui and Meng [69]; Wright et al. [70].

The application of P at the rate of 74 kg P_2O_5 ha^{-1} significantly increased seed index (weight of 100 seed in g) relative to the application at 44 kg P_2O_5 ha^{-1} (Table 7) [36]. A possible explanation for increased seed weight due to the application of P at the higher rate is that this nutrient activated biological reactions in the cotton plants, particularly CO_2 fixation and the synthesis of sugar, amino acids, protein, lipids and other organic compounds. It also increased the translocation of assimilates from photosynthetic organs to the sink [71]. Similar results were obtained by El-Debaby et al. [72]. Application of Zn significantly increased seed index, compared to the control [36]. This may be due to its favorable effect on photosynthetic activity. Zinc improves mobilization of photosynthates and directly influences boll weight that coincide directly with increased seed index. These results are confirmed by those obtained by Ibrahim et al. [13]. Calcium applied at all rates significantly increased seed index over the control [36]. The highest rate of Ca (60 ppm) showed the highest numerical value of seed index. Similar results were obtained by Ibrahim et al. [13].

Raising the phosphorus rate increased seed oil content and oil yield ha^{-1} (Table 7) [36]. This may be attributed to the fact that P is required for production of high quality seed, since it occurs in coenzymes involved in energy transfer reactions. Energy is tapped in photosynthesis in the form of adenosine triphosphate (ATP) and nicotinamide adenine dinucleotide phosphate (NADP). This energy is then used in photosynthetic fixation of CO_2 and in the synthesis of lipids and other essential organic compounds [8]. These results agree with those obtained by Pandrangi et al. [55]; Gebaly and El-Gabiery [14]. Spraying plants with zinc resulted in an increase of seed oil content and oil yield ha^{-1} when compared with the untreated control [36]. This could be attributed to the increase of total photo assimilates (e.g. lipids) and the translocated assimilates to the sink as a result of applying zinc. Similar results were reached by Ibrahim et al. [13]. Application of Ca at all concentration tended to increase the seed oil content and oil yield ha^{-1} over the control; the best result was from the highest Ca concentration (60 ppm) [36]. These results agreed with those obtained by Bora [73] on rape; Ibrahim et al. [13] on cotton; Bybordi and Mamedov [59] on canola. A possible role of Ca as an activator of the enzyme phospholipase in cabbage leaves has been investigated by Davidson and Long [74].

Applying P at the higher rate slightly increased seed protein content (Table 7) [36]. It also increased the protein yield ha^{-1}, resulting from an improvement in both seed yield and seed protein content. Phosphorus is a component of nucleic acids which are necessary for protein synthesis [66]. Similar results were obtained by Ibrahim et al. [13]; Gebaly and El-Gabiery [14] in cotton. The application of Zn increased the seed protein content and protein yield ha^{-1}, compared with the untreated control [36]. Shchitaeva [75] found that the synthesis of metabolically active amino acids depends on Zn, which increases the synthesis of asparagine and tryptophan. These results agree with studies reported by Ibrahim et al. [13]. Calcium applied at all rates tended to decrease the seed protein content slightly, but protein yield ha^{-1} increased compared with the untreated control, which is attributed to the increase in cottonseed yield. The best protein yield was obtained at the highest Ca concentration (60 ppm) [36].

The oil refractive index and unsaponifiable matter tended to increase, while the acid value and saponification value tended to decrease as phosphorus rate was raised (Table 8) [36]. The increase in unsaponifiable matter is known to be beneficial, as it increases oil stability. Spraying plants with Zn resulted in a slight increase in the oil refractive index and unsaponifiable matter and a slight decrease in acid value and saponification value, compared with the untreated control [36]. Similar results were obtained by Sawan et al. [57] concerning the effect of applied Zn on oil refractive index, unsaponifiable matter and saponification value. Application of Ca at any concentration tended to decrease the oil acid value and saponification value and to increase the unsaponifiable matter, especially as the applied Ca concentration increased, compared with the untreated control [36]. This became especially apparent as the applied calcium concentration was increased. The effect of Ca concentrations on oil refractive index was very limited and without a defined trend. These results are in agreement with those reported by Sawan et al. [57] concerning the effect of applied Ca on oil refractive index, saponification value and unsaponifiable matter. The studied oil quality characters seemed to be genetically controlled.

The high rate of applied P decreased the oil saturated fatty acids capric, myristic, palmitic and stearic, while it increased lauric acid (Table 9). The total saturated fatty acids also decreased. Palmitic acid was the predominant saturated fatty acid. Low content of saturated fatty acids is desirable for edible uses [36]. The application of Zn decreased the abundant saturated fatty acids palmitic and myristic, while it increased capric, lauric and stearic saturated fatty acids, compared to the control. The total saturated fatty acids decreased [36].

Table 8: Effect of P rate and foliar application of Zn and Ca on seed oil properties[a]

Treatments		Refractive index	Acid value	Saponification value	Unsaponifiable matter (%)
P2O5 rate (ktha-1)					
Control	44	1.4688	0.1332	192.9	0.3575
	74	1.4691	0.1327	191.5	0.3662
S.E[b]		0.0001	0.0002	0.7	0.0043
Zn rate (ppm)					
Control	0	1.4687	0.1331	192.4	0.3538
	40	1.4692	0.1328	192.0	0.3700
S.E[b]		0.0002	0.0001	0.2	0.0081
Ca rate (ppm)					
Control	0	1.4689	1.1340	194.9	0.3550
	20	1.4688	0.1330	191.4	0.3575
	40	1.4688	0.1324	191.3	0.3650
	60	1.4692	0.1324	191.2	0.3700
S.E[b]		0.0001	0.0003	0.9	0.0034

[a]Mean data from a four replicate composite for the two seasons. [b]S.E. = standard error [36].

Table 9: Effect of P rate and foliar application of Zn and Ca on the relative percentage of saturated fatty acids[a]

Treatments			Relative % of saturated fatty acids					
			Capric	Laurie	Myristic	Palmitic	Stearic	Total
P2O5 rate (kg ha-1)								
Control		44	0.0812	0.1212	0.5100	21.65	1.844	24.206
		74	0.0688	0.1538	0.2612	19.95	1.746	22.180
8.0			0.0062	0.0163	0.1244	0.85	0.049	1.013
Zn rate (ppm)								
Control		0	0.0500	0.0988	0.3912	21.67	1.752	23.962
		40	0.1000	0.1762	0.3800	19.93	1.838	22.424
S.Eb			0.0250	0.0387	0.0056	0.87	0.043	0.769
Ca rate (ppm)								
Control		0	0.1375	0.2575	0.5025	22.36	2.090	25.347
		20	0.0600	0.1100	0.2900	21.15	1.742	23.352
		40	0.0300	0.0825	0.3000	19.59	1.090	21.092
		60	0.0725	0.1000	0.4500	20.09	2.258	22.970
S.Eb			0.0226	0.0404	0.0534	0.61	0.258	0.872

[a]Mean data from a four replicate composite for the two seasons. [b]S.E. = standard error [36].

Calcium applied at all concentrations decreased in the saturated fatty acids capric, lauric, myristic, palmitic and stearica as well as the saturated fatty acids compared with the untreated control with one exception [36]. Spraying plants with Ca at 60 ppm tended to increase stearic acid, compared with the control. Applied Ca at 40 ppm gave the lowest capric, lauric, palmitic and stearic acids contents, compared with the other two concentrations (20 and 60 ppm). Calcium at 20 ppm gave the lowest myristic acid content, compared with 40 and 60 ppm. The total unsaturated fatty acids (oleic and linoleic) and the ratio between total unsaturated fatty acids and total saturated fatty acids (TU/TS) were increased by raising P rate (Table 10). Linoleic acid was the most abundant unsaturated fatty acid. Gushevilov and Palaveeva [76] studied the changes in sunflower oil contents of linoleic, oleic, stearic and palmitic acids due to application rate of phosphorus and found that oil quality remained high at a high P rate. The application of Zn resulted in an increase in total unsaturated fatty acids and TU/

TS ratio, over the control [36]. Calcium applied at all rates increased the total unsaturated fatty acid and TU/TS ratio, compared with untreated control. Calcium at 40 ppm gave the highest increment, total unsaturated fatty acid and TU/TS ratio, followed by 60 ppm concentration. Spraying plants with Ca at 20 ppm produced seed oil characterized by the highest oleic acid content, while spraying with 40 ppm gave the highest linoleic acid content, compared with the other concentrations [36].

Cottonseed, Protein, Oil Yields, and Oil Properties as Influenced by K, P and Zn

Seed yield ha^{-1} significantly increased when K was applied (by as much as 13.99%) (Table 11) [1]. Potassium would have a favorable impact on yield components, including a number of open bolls plant^{-1} and boll weight, leading to a higher cotton yield. The role of K suggests that it affects abscission (reduced boll shedding) and it certainly affects yield [53]. Gormus [17]; Ibrahim et al. [13]; Gebaly [77] also found that K application increased yield. Application of Zn significantly increased seed yield ha^{-1}, as compared with the untreated control (by 9.38%) [1]. A possible explanation of such results might be the improvement of yield components due to the application of Zn.

Table 10: Effect of P rate and foliar application of Zn and Ca on the relative percentage of unsaturated fatty acids[a]

Treatments			Relative % of unsaturated fatty acids			TU/TSb°
			Oleic	Linoleic	Total	ratio
P2O5 rate (kg ha-1)						
Control		44	21.89	53.90	75.79	3.13
		74	21.91	55.91	77.82	3.51
S.Ec			0.01	1.00	1.01	0.19
Zn rate (ppm)						
Control		0	21.70	54.33	76.03	3.17
		40	22.09	55.48	77.57	3.46

S.Ec		0.19	0.57	0.77	0.14
Ca rate (ppm)					
Control	0	21.34	53.31	74.65	2.94
	20	22.26	54.38	76.64	3.28
	40	22.00	56.90	78.90	3.74
	60	22.00	55.02	77.02	3.35
S.Ec		0.19	0.75	0.87	0.16

[a]Mean data from a four replicate composite for the two seasons. [b]TU/TS ratio = (total unsaturated fatty acids)/(total saturated fatty acids). [c]S.E. = standard error [36].

Table 11: Effect of K rate and foliar application of Zn and foliar, additional P on cottonseed yield, seed index, seed oil, seed protein, oil and protein yields

Treatments	Cottonseed yield (kg. ha-1)a	Seed index (g)a	Seed Oil (%)b	Oil yield (kg ha-1)b	Seed protein (%)b	Protein yield (kg-ha-l)b
K rate (kg• ha-l)						
0, control	1828.0	10.01	19.55	357.5	22.24	406.6
47.4	2083.8	10.16	19.82	413.2	22.27	464.1
L.S.D. 0.05c	80.6	0.05				
S.D.c			0.15	34.2	0.03	36.2
Zn rate (g ha-l)						
0, control	1868.3	10.04	19.59	366.2	2225	415.7
57.6	2043.5	10.13	19.78	404.4	2226	455.0
L.S.D. 0.05c	80.6	0.05				
S.D.c			0.18	40.5	0.04	42.6
P rate (3• ha-l)						
0, control	1775.8	9.97	19.56	347.5	2223	394.8
576	1944.3	10.08	19.64	382.1	2225	432.7
1152	2023.7	10.13	19.76	400.3	22.26	450.5
1728	2079.8	10.16	19.77	411.5	22.28	463.3

L.S.D. 0.05c	114.0	0.07				
S.D.c			0.20	40.2	0.04	41.7

[a]Combined statistical analysis from the two seasons. [b]Mean data from a four replicate composites for the two seasons. [c]L.S.D. = least significant differences. [d]S.D. = standard deviation was used to conduct t-test to verify the significance between every two treatment means at 0.05 level [1].

Zinc could have a favorable effect on photosynthetic activity of leaves [22], which improves mobilization of photosynthates and directly influences boll weight. Further, Zn is required in the synthesis of tryptophan, a precursor of indole-3-acetic acid [23], which is the major hormone, inhibits abscission of squares and bolls. Thus the number of retained bolls plant^{-1} and consequently seed yield ha^{-1} would be increased [24]. Similar results were obtained by Ibrahim et al. [13]. Phosphorus extra foliar application at all the three concentrations (576, 1152 and 1728 g of P ha^{-1}) also significantly increased seed yield ha^{-1}, where the three concentrations applied proved to excel the control (by 9.49% - 17.12%). The best yield was obtained at the highest P concentration tested [1]. Such results reflect the pronounced improvement of yield components due to application of P which is possibly ascribed to its involvement in photosynthesis and translocation of carbohydrates to young bolls [9]. Phosphorus as a constituent of cell nucleus is essential for cell division and the development of meristematic tissue and hence it would have a stimulating effect on increasing the number of flowers and bolls plant^{-1} [78]. These results agree with that reported by Ibrahim et al. [13]; Saleem et al. [67]; Gebaly and El-Gabiery [14].

Seed index significantly increased with applying K (Table 11) [1]. A possible explanation for the increased seed index due to the application of K may be due in part to its favorable effects on photosynthetic activity rate of crop leaves and CO_2 assimilation [15], which improves mobilization of photosynthates and directly influences boll weight which in turn directly affects seed weight [79]. The application of Zn significantly increased seed index, as compared to control [1]. The increased seed weight might be due to an increased photosynthesis activity resulting from the application of Zn [22] which improves mobilization of photosynthates and the amount of photosynthate available for reproductive sinks and thereby influences boll weight, factors that coincide with increased in seed weight [24]. The phosphorus applied at all three rates significantly increased seed index over the control. The highest rate of P (1728 g·ha^{-1}) showed the highest numerical value of seed index [1]. This increased seed weight may be due to the fact that P activated the biological reaction in cotton plant, particularly photosynthesis fixation of CO_2 and synthesis of sugar, and other organic compounds [22,80]. This indicates that treated cotton bolls had larger photosynthetically supplied sinks for carbohydrates and other metabolites than untreated bolls [1].

The applied K caused significant increase in seed oil content and oil yield ha^{-1} (55.7 kg oil ha^{-1}), compared with untreated control (Table 11) [1]. This could be attributed to the role of K in biochemical pathways in plants. Potassium increases the photosynthetic rates of crop leaves, CO_2 assimilation and facilitates carbon movement [15]. The favorable effects of K on seed oil content and oil yield were mentioned by Ibrahim et al. [13]; Gebaly [77]. Spraying plants with Zn resulted in an increase in seed oil content and oil yield ha^{-1} (38.2 kg oil ha^{-1}), compared with the untreated control. Cakmak [81] has speculated that Zn deficiency stress may inhibit some antioxidant enzymes, resulting in extensive oxidative damage to membrane lipids. Similar results were obtained by Ibrahim et al. [13]. The foliar application of P at all the three concentrations tended to increase the seed oil content and oil yield ha^{-1} (34.6 - 64.0 kg oil ha^{-1}), over the control [1]. The effect was the most significant at the highest P concentration (1728 g·ha^{-1}) on oil yield ha^{-1}. These results agree with those obtained by Ibrahim et al. [13]; Gebaly and El-Gabiery [14].

The applied K caused a slight increase in seed protein content and significantly increased protein yield ha^{-1} (57.5 kg protein ha^{-1}), compared with the untreated control (Table 11) [1]. It also increased the protein yield ha^{-1}, resulting in an improvement in both seed yield and seed protein content. This could be attributed to the role of K in biochemical pathways in plants. Potassium increases the photosynthetic rates of crop leaves, CO_2 assimilation and facilitates carbon movement [15]. Also, K has favorable effects on metabolism of nucleic acids, and proteins [82]. These are manifested in metabolites formed in plant tissues and directly influence the growth and development processes. Similar results were obtained by Ibrahim et al. [13]; Gebaly [77]. The application of Zn slightly increased the seed protein content, and increased protein yield ha^{-1} (39.3 kg protein ha^{-1}) numerically compared with the untreated control. Because Zn is directly involved in both gene expression and protein synthesis. Cakmak [81] has speculated that Zn deficiency stress may inhibit the activities of a number of antioxidant enzymes, resulting in extensive oxidative damage to proteins, chlorophyll and nucleic acids. These results agree with those reported by Ghourab et al. [79]. Phosphorus applied at all rates tended to increase the seed protein content and the protein yield ha^{-1} (37.9-68.5 kg protein ha^{-1}) compared with the untreated control [1]. The effect was significant on protein yield ha^{-1} when applied the high P concentration (1728 g·ha^{-1}), resulting from an improvement in both seed yield and seed protein content. Phosphorus is a component of nucleic acids, which are necessary for protein synthesis [8]. Similar results were obtained by Gebaly and El-Gabiery [14].

The oil refractive index, unsaponifiable matter and iodine value significantly increased, while saponification value significantly decreased by applied K, compared with the untreated control (Table 12) [1]. On the other hand, the acid value was not significantly affected due to the K application.

Table 12: Effect of K rate and foliar application of Zn and foliar, additional P on seed oil properties[a]

Treatments	Refractive index	Acid value	Saponification value	Unsaponifiable matter (%)	Iodine value
K rate (ktha-l)					
0, control	1.4684	0.1343	190.81	0.3538	127.48
47.4	1.4698	0.1316	189.74	0.3950	132.76
S.Db	0.0013	0.0032	0.74	0.0223	3.63
Zn rate (g ha-1)					
0, control	1.4683	0.1336	190.71	0.3625	128.39
57.6	1.4699	0.1323	189.84	0.3863	131.85
S.Db	0.0012	0.0034	0.80	0.0287	4.21
P rate (g.ha-l)					
0, control	1.4681	0.1350	190.75	0.3525	125.33
576	1.4693	0.1343	190.33	0.3725	131.46
1152	1.4696	0.1323	190.10	0.3800	131.93
1728	1.4695	0.1309	189.92	0.3925	131.76

| S.Db | 0.0015 | 0.0033 | 0.94 | 0.0294 | 3.80 |

[a]Mean data from a four replicate composites for the two seasons. [b]S.D. = standard deviation [1]

Potassium is an essential nutrient and an integral component of several important compounds in plant cells. This attributed to the role of K in biochemical pathways in plants [83]. These may be reflected in distinct changes in seed oil quality. Mekki et al. [84] stated that, foliar application with K (0 or 3.5% K_2O) on sunflower at the seed-filling stage resulted in decreased oil acid content. Froment et al. [85], in linseed, found that the iodine value, which indicates the degree of unsaturation in the final oil, was highest in treatments receiving extra K. Spraying plants with Zn resulted in a significant increase in oil refractive index, and a significant decrease in unsaponifiable matter, compared with untreated control. The other oil properties (acid, saponification, and iodine values) were not significantly affected. Zinc activates a large number of enzymes, either due to binding enzymes and substrates, or the effects of Zn on conformation of enzymes or substrate, or both [86,87]. These would have a direct impact through utilization in the growth processes, which are reflected in distinct changes in seed oil quality [1]. The application of P at all concentrations significantly increased iodine value, compared with the untreated control, while the other oil properties (oil refractive index; acid and saponification values, and the unsaponifiable matter) were not significantly affected.

The applied K decreased the oil-saturated fatty acids (capric, lauric, myristic, palmitic, and stearic) (Table 13) [1]. A significant effect was found only on capric, palmitic, and the total saturated fatty acids. The total unsaturated fatty acids (oleic and linoleic) and the ratio between total unsaturated fatty acids and total saturated fatty acids (TU/TS) were increased (by 4.31, and 19.77%, respectively) by applied K (Table 14) [1]. The effect was significant on linoleic acid, the total unsaturated fatty acids (oleic and linoleic), and TU/TS ratio. The beneficial effect of applied K on TU and TU/TS ratio may be due to the regulated effect of K, which acts as an activator on many enzymatic processes, where some of these enzymes may affect the seed oil content from these organic matters. To our knowledge, no information on the effect of K on the cottonseed oil fatty acids is available in the literatures [1]. Mekki et al. [84] stated that, foliar application with K on sunflower increased the oleic acid fatty acid. Froment et al. [85], in linseed oil, found that the linoleic acid content was greatest in treatment receiving extra K. The application of Zn resulted in a decrease of the saturated fatty acids, i.e. palmitic, capric, myristic, and stearic, and the total, but resulted in an increase in lauric acid, compared to the untreated control [1]. The effect was significant only on palmitic acid, and the total saturated fatty acids in the oil. The application of Zn resulted in an increase in the total unsaturated fatty acids (by 3.49%) and TU/TS ratio (by 15.25%), over the control. The effect was significant on oleic acid, the total unsaturated fatty acids (oleic and linoleic), and TU/TS ratio. The stimulatory residual effects of the application Zn on TU and TU/TS ratio were probably due to the favorable effects of Zn on fundamental metabolic reactions in plant tissues. Phosphorus applied at all concentrations resulted in a decrease in the total saturated fatty acids compared with the untreated control. Spraying plants with P at 1728 g·ha^{-1} gave the lowest total saturated fatty acids oil, followed by P at 1152 g·ha^{-1} concentration, compared with the control [1]. Application the high P concentration (1728 g·ha^{-1}) gave the lowest capric, lauric, palmitic, and stearic acid contents compared with the other two concentrations (576 and 1152 g of P ha^{-1}), while applied P at 1152 g·ha^{-1} gave the lowest myristic acid content compared with the other two concentrations (576 and 1728 g of P ha^{-1}).

Table 13: Effect of K rate and foliar application of Zn and foliar, additional P on the relative percentage of saturated fatty acids[a]

Treatments	Relative % of saturated fatty acids					
	Capric	Lauric	Myristic	Palmitic	Stearic	Total
K rate (kg. ha-l)						
0, control	0.0774	0.0626	0.8275	22.21	2.271	25.452
47.4	0.0728	0.0599	0.4863	19.72	1.915	22.250
S. Db	0.0036	0.0079	0.3407	1.48	0.451	2.331
Zn rate (g ha-1)						
0, control	0.0769	0.0609	0.6763	22.16	2.185	25.159
57.6	0.0733	0.0616	0.6375	19.77	2.001	22.544
S.Db	0.0040	0.0049	0.3859	1.79	0.479	2.532
P rate (g• ha-l)						
0, control	0.0795	0.0665	1.1075	22.80	2.728	26.776
576	0.0748	0.0623	0.5925	20.70	1.855	23.287

1152	0.0733	0.0595	0.4375	20.30	1.905	22.770
1728	0.0728	0.0568	0.4900	20.07	1.885	22.572
S.Db	0.0036	0.0034	0.2826	2.02	0.317	2.422

[a]Mean data from a four replicate composites for the two seasons. [b]S.D. = standard deviation [1].

Table 14: Effect of K rate and foliar application of Zn and foliar, additional P on the relative percentage of unsaturated fatty acids[a]

Treatments	Relative % of unsaturated fatty acids			TU/TS[b] ratio
	Oleic	Linoleic	Total	
K rate (kg ha-')				
0, control	21.61	52.94	74.54	2.954
47.4	22.73	55.01	77.75	3.538
S.D[c]	1.40	1.49	2.33	0.403
Zn rate (g. ha-1)				
0, control	21.43	53.40	74.84	3.016
57.6	22.90	54.55	77.45	3.476
S.D[c]	1.31	1.76	2.53	0.446
P rate (t ha-1)				
0, control	21.11	52.11	73.22	2.755
576	21.96	54.75	76.70	3.331
1152	22.52	54.70	77.23	3.427
1728	23.09	54.33	77.43	3.472
S.D[c]	1.42	1.57	2.42	0.439

[a]Mean data from a four replicate composite for the two seasons. [b]TU/TS ratio = (total unsaturated fatty acids)/(total saturated fatty acids). [c]S.D. = standard deviation [1].

The effect was significant for the two concentrations 1152 and 1728 g of P ha^{-1} on capric acid and the total saturated fatty acids in the oil, and for all different P concentrations on lauric, myristic, and stearic. Phosphorus applied at all rates increased the total unsaturated fatty acid (by 4.77% - 5.75%) and TU/TS ratio (by 20.91% - 26.03%) compared with the untreated control. Applied P at 1728 g·ha^{-1} gave the highest increment, followed by the concentration 1152 g of P ha^{-1} [1]. Spraying plants with P at 1728 g·ha^{-1} produced seed oil characterized by the highest oleic acid content, while spraying with 576 g of P ha^{-1} gave the highest linoleic acid content compared with the other concentrations. The effect was significant for the high P concentration (1728 g·ha^{-1}) on oleic, for the two concentrations, i.e., 1152 and 1728 g of P ha^{-1} on the TU/TS ratio, and for all different concentrations on linoleic, and the total unsaturated fatty acid [1]. The beneficial effect of applied P at different concentrations on TU and TU/TS ratio may be due to the regulated effect of P on many enzymatic processes and the fact that P acts as an activator of some enzymes which may affect the seed oil fatty acids composition. Gushevilov and Palaveeva [76] studied the changes in the contents of linoleic, oleic, stearic, and palmitic acids in sunflower oil due to the P-application rate and found that oil quality remained high at a high P-rate. Khan et al. [88] indicated that oleic acid increased by increasing levels of P added to rapeseed mustard.

Effects N, K and PGR on Oil Content and Quality of Cotton Seed

The seed yield of cotton significantly ($P < 0.05$) increased (as much as 13.03%) by increasing N-application rate from 95.2 to 142.8 kg·ha^{-1} (Table 15) [37]. There is an optimal relationship between the nitrogen content in the plant and CO_2 assimilation, where decreases in CO_2 fixation are well documented for N-deficient plants. Nitrogen deficiency is associated with elevated levels of ethylene (which increase boll shedding), suggesting ethylene production in response to N-deficiency stress [89]. Nitrogen is also an essential nutrient in creating plant dry matter, as well as many energy-rich compounds which regulate photosynthesis and plant production, thus influencing boll development, increasing the number of bolls per plant and boll weight. Similar findings were obtained by McConnell and Mozaffari [90] and Saleem et al. [46] when N fertilizer was applied at 120 kg·ha^{-1} and Wiatrak et al. [80] when N fertilizer was applied at 67 - 202 kg·ha^{-1}. Also, similar results were obtained by Sarwar Cheema et al. [33]; Hamed et al. [47]. On the other hand Boquet (2005) reported that increasing N from 90 to 157 kg·ha^{-1} did not result in increased cotton yield in irrigated or rain-fed cotton. Foliar application of K significantly increased seed yield by 10.02% to 16.25% as compared to the control (0 g·K·ha^{-1}) (Table 15) [37]. The differences between the effects of the three concerned K rates were statistically insignificant; with the exception of the 957 g·K·ha^{-1} concentration that proved to produce significantly higher seed yield ha^{-1} (5.66%) than the 319 g·K·ha^{-1} concentration. These increases could be due to the favorable effects of this nutrient on yield components such as number of opened bolls plant^{-1}, boll weight, or both, leading to higher cotton yield. Zeng [53] indicated that, K fertilizer reduced boll shedding. Pettigrew [91] stated that, the elevated carbohydrate concentrations remaining in source tissue, such as leaves, appear to be part of the overall effect of K deficiency in reducing the amount of photosynthate available for reproductive sinks and thereby producing changes in boll weight. Cakmak et al. [92] found that, the K nutrition had pronounced effects on carbohydrate partitioning by affecting either the phloem export of photosynthates (sucrose) or growth rate of sink and/or source organs. Mullins et al. [93] evaluated cotton yield under a long-term soil application of K at 75 - 225 kg K$_2$O ha^{-1}, and found that K application increased yield. Results obtained here confirmed those obtained by Aneela et al. [18] when applying 200 kg K$_2$O ha^{-1}, Pervez et al. [19] under 62.5, 125, 250 kg·K·ha·ha^{-1}, Pettigrew et al. [20] under K fertilizer (112 kg·ha^{-1}); Gebaly [77]. Application of the PGR mepiquat chloride significantly increased seed yield ha^{-1} (by 9.72%), as compared with untreated plants.

Table 15: Effect of soil application of N and foliar application of K and mepiquat chloride (MC) on the yield, 100-seed weight, oil and protein of the cotton

Treatments	Cottonseed yield (kg. ha-1)a	100-seed weight (g)a	Seed oil (%)b	Oil yield (kg. ha-1)b	Seed protein (%)b	Protein yield (kg.ha)b
N rate (kg.ha-1)						
95.2	1862.4	10.09	19.73	367.5	22.24	414.2
142.8	2105.0	10.32	19.60	413.0	22.44	472.2
L.S.D. 0.05c	78.7	0.07				-
S.D.c		-	0.16	33.6	0.11	35.5
K rate (g ha-l)						

0	1804.4	10.03	19.49	351.6	22.32	402.9
319	1985.2	10.19	19.61	389.3	22.32	443.1
638	2047.7	10.27	19.73	404.2	22.34	457.7
957	2097.6	10.32	19.83	415.8	22.37	469.3
L.S.D. 0.05c	111.4	0.10				
S.D.c		-	0.12	35.0	0.16	41.8
MC rate (g ha-1)						
0	1891.8	10.13	19.61	371.1	22.31	422.1
48 + 24	2075.6	10.27	19.72	409.4	22.37	464.4
L.S.D. 0.05c	78.7	0.075				-
S.D.d		-	0.17	36.1	0.15	41.3

[a]Combined statistical analysis from the two seasons. [b]Mean data from a four replicate composites for the two seasons. [c]L.S.D. = least significant differences, [d]S.D. = standard deviation was used to conduct t-test to verify the significance between every two treatment means at 0.05 level [37].

Such increases could be due to the fact that, the application of mepiquat chloride restrict vegetative growth and thus enhance reproductive organs by allowing plants to direct more energy towards the reproductive structure [51]. This means that bolls on treated cotton would have a larger photo synthetically supplied sink of carbohydrates and other metabolites than did those on untreated cotton [30]. Results agreed with those obtained by Prakash et al. [94] when mepiquat chloride was applied at 50 ppm, Mekki [95] when mepiquat chloride was applied at 100 ppm, and Kumar et al. [31]. Also, similar results were obtained by Sarwar Cheema et al. [33]; Gebaly and El-Gabiery [14].

Seed weight significantly increased by adding the high N-rate (Table 15) [37]. This may be due to increased photosynthetic activity that increases accumulation of metabolites, with direct impact on seed weight. Reddy et al. [2], in a pot experiment under natural environmental conditions, where 20-day old cotton plants received 0, 0.5, 1.5 or 6 mM NO_3, found that, net photosynthetic rates, stomatal conductance and transpiration were positively correlated with leaf N concentration. Similar findings were reported by Palomo et al. [96], when N was applied at 40 - 200 kg·ha^{-1}; Ali and El-Sayed [97], when

N was applied at 95 to 190 kg·ha^{-1}; Hamed et al. [47] when N was applied up to 178 kg·ha^{-1}. 100-seed weight significantly increased with K application at all the three concentrations as compared to the control [37]. The highest rate of K (957 g·K·ha^{-1}) resulted the highest seed weight. The difference between the high rate and low rate (319 g·K·ha^{-1}) was also significant. Increase in seed weight might be due to the effect of K on mobilization of photosynthates, which would directly influence boll weight and increase seed weight [38,91]. Ibrahim et al. [13] reported that, the application of K fertilizer resulted in an increase in seed weight. The application of mepiquat chloride significantly increased 100-seed weight as compared to the plots that had not received mepiquat chloride, the untreated control [37]. Increased seed weight as a result of mepiquat chloride applications may be due to an increase in photosynthetic activity, which stimulates photosynthetic activity, and dry matter accumulation [31,82], and in turn increases the formation of fully-mature seed and thus increases seed weight. Similar results to the present study were obtained by Ghourab et al. [79]; Lamas [98].

Seed oil content was slightly decreased with an increase in the N rate from 95.2 to 142.8 kg·ha^{-1}, but seed oil yield ha^{-1} had significantly increased (45.5 kg oil ha^{-1}), which is attributed to the significant increase in seed yield (Table 15) [37]. Similar results were obtained by Froment et al. [85], in linseed; Zubillaga et al. [99] in sunflower. Yield increases in this study were attributed to the fact that N was an important nutrient in controlling new growth, thus influencing boll development, increasing the number of bolls plant^{-1} and boll weight. Synthesis of fat requires both N and carbon skeletons during the course of seed development [5]. The application of K at all the three concentrations tended to increase seed oil content and yield over the control (37.7 - 64.2 kg oil ha^{-1}), but was statistically significant only for 638 and 957 g·K·ha^{-1} concentrations on the seed oil content, and with K application at all the three concentrations on the oil yield ha^{-1} [37]. The highest rate of K (957 g·K·ha^{-1}) showed the highest numerical values of seed oil content and oil yield ha^{-1} compared with the other two concentrations (319 and 638 g·K·ha^{-1}) [37]. This could be attributed to the role of K in biochemical pathways in plants. Pettigrew [91] stated that, the elevated carbohydrate concentrations remaining in source tissue, such as leaves, appear to be part of the overall effect of K deficiency in reducing the amount of photosynthate available for reproductive sinks and thereby producing changes in yield and quality found in cotton. Madraimov [100] indicated that, increasing the rates of applied K_2O from 0 to 150 kg·ha^{-1} produced linear increases in cottonseed oil contents. Previously, favorable effects of K on seed oil content and oil yield were mentioned by Ibrahim et al. [13]; Gebaly [77]. They reported that, increasing K supply to maternal cotton plants increased crude fat content of seed. The application of mepiquat chloride resulted in an insignificant increase in seed oil content over that of the control [37]. Also significantly increased the seed oil yield ha^{-1} compared with the untreated control (by 38.3 kg oil ha^{-1}). These results could be attributed to the increase of total photoassimilates (e.g. lipids) and the translocated assimilates to the sink as a result of applying mepiquat chloride [101]. This result agreed with those obtained by Gebaly and El-Gabiery [14].

High N-rate significantly increased the seed protein content and yields (58.0 kg protein ha^{-1}) (Table 15) [37]. Stitt [102] indicated that, nitrate (NO_3^-) induces genes involved in different aspects of carbon metabolism, including the synthesis of organic acids used for amino acid synthesis. These results suggest that the highest N rate of the added N in this study compared with the lowest rate increases the amino acids synthesis in the leaves and this stimulate the accumulation of protein in the seed. The present results confirmed the findings of Patil et al. [34]. Averaged seed protein content tended to increase when applying 638 and 957 g·K·ha^{-1} compared with untreated control (0 g·K·ha^{-1}) [37]. Applied K at all rates also, increased the protein yield numerically (40.2 - 66.4 kg protein ha^{-1}), resulting from an improvement in both seed yield and seed protein content. The increase in protein yield ha^{-1} was statistically significant when applying the 638 and 957 g·K·ha^{-1} concentrations. Best protein yield was obtained at the high K concentration (957 g·K·ha^{-1}) compared with the other two concentrations (319 and 638 g·K·ha^{-1}) [37]. This could be attributed to the role of K in biochemical pathways in plants. Potassium has favorable effects on metabolism of nucleic acids and proteins [82]. These are manifested in metabolites formed in plant tissues, and directly influence the growth and development processes, thereby producing changes in yield and quality of cotton [37]. These results were in agreement with those obtained by Ibrahim et al. [13]; Gebaly [77]. Seed protein content tended to increase numerically, while seed protein yield was significantly increased (42.3 kg protein ha^{-1}) in plants treated with mepiquat chloride as compared with the untreated plants. The increase in seed protein content and yield may be caused by the role of mepiquat chloride in protein synthesis, encouraging the conversion of amino acids into protein [62] along with the favorable and significant effect of mepiquat chloride on cottonseed yield. These results were confirmed by Gebaly and El-Gabiery [14].

The seed oil refractive index, unsaponifiable matter and iodine value tended to increase, while the oil saponification and acid values tended to decrease by raising N-rate (Table 16) [37]. Narang et al. [103] indicated that, N application increased the oil-quality index (iodine number) in rape. The application of K at different concentrations tended to increase the seed oil refractive index, unsaponifiable matter and iodine value, and to decrease the oil saponification value and acid value, numerically, compared with the untreated control, especially when applied K at the high concentration (957 g·K·ha^{-1}) [37]. The effect was significant for the two concentrations 638 and 957 g·K·ha^{-1} on acid value, and unsaponifiable matter, and for all different concentrations on iodine value. The effect of K concentrations on oil refractive index was very limited. Potassium is an essential nutrient and an integral component of several important compounds in plant cells. This attributed to the role of K in biochemical pathways in plants, where K acts as an activator for several enzymes involved in carbohydrates metabolism [8]. These may be reflected in distinct changes in seed oil quality [37]. Mekki et al. [84] stated that, foliar application with K (0 or 3.5% K$_2$O) on sunflower at the seed-filling stage, decreased oil acid value. Froment et al. [85], when working with linseed found that, the iodine value, which indicates the degree of unsaturation of the final oil, was highest in treatment receiving extra K. The application of mepiquat chloride tended to significantly increase the oil refractive index, unsaponifiable matter and iodine value, while it tended to insignificantly decrease the oil acid value and saponification value, compared with the untreated control [37]. The application of plant growth regulators, particularly growth retardants may maintain internal hormonal balance, and efficient sink source relationship. This may be reflected in distinct changes in seed oil quality.

Saturated fatty acids in oil, lauric, myristic, palmitic and their total decreased, while capric and stearic increased by raising the N-rate (Table 17) [37]. The effect was significant only on palmitic acid, which was the dominant saturated fatty acid. A low content of saturated fatty acids is desirable for edible. The total unsaturated fatty acids (oleic and linoleic) and the ratio between total unsaturated fatty acids and total saturated fatty acids (TU/TS) were increased (by 2.42, and 10.69%, respectively) by raising N-rate (Table 18) [37]. The effect was significant only on oleic acid. Linoleic acid was the most abundant unsaturated fatty acid. Holmes and Bennett [104] commented that, the fatty acid composition of rape oil is mainly under genetic control, but can be modified to some extent by N nutrition.

Table 16: Effect of N rate and foliar application of K and mepiquat chloride (MC) on seed oil properties[a]

Treatments	Refractive index	Acid value	Saponification value	Unsaponifiable matter (%)	Iodine value
N rate (ktha-1)					
95.2	1.4684	0.1339	190.8	0.3762	128.9
142.8	1.4695	0.1313	189.7	0.3913	131.1
S.D.[b]	0.0011	0.0025	1.4	0.0178	33
K rate (g.ha-l)					
0	1.4682	0.1352	190.8	0.3675	125.8
319	1.4689	0.1337	190.1	0.3825	130.3
638	1.4692	0.1315	190.3	0.3875	131.6
957	1.4694	0.1300	190.1	0.3975	132.4
S.D.[b]	0.0012	0.0021	1.5	0.0170	2.5
MC rate (g.ha-1)					
0	1.4683	0.1331	190.6	0.3750	128.3
48 + 24	1.4696	0.1321	189.9	0.3925	131.7
S.D.[b]	0.0011	0.0028	1.6	0.0172	3.0

[a]Mean data from a four replicate composites for the two seasons. [b]S.D. = standard deviation [37].

Table 17: Effect of N rate and foliar application of K and mepiquat chloride (MC) on the relative percentage of saturated fatty acids[a]

Treatments	Relative % of saturated fatty acids					
	Capric	Lauric	Myristic	Palmitic	Stearic	Total
N rate (kg.ha-l)						
95.2	0.068	0.068	0.691	21.77	2.157	24.753
142.8	0.069	0.067	0.645	20.18	2.969	22.934
S.D.c	0.009	0.006	0.451	1.44	0.470	2.283
K rate (g. ha-l)						
0	0.077	0.074	1.307	22.40	2.602	26.467
319	0.072	0.070	0.675	21.02	1.955	23.792
638	0.065	0.063	0.350	20.52	1.905	22.903
957	0.061	0.062	0.340	19.96	1.790	22.212
S.D.c	0.006	0.004	0.180	1.47	0.369	1.925
MC rate (g• ha-l)						
0	0.074	0.065	0.775	21.97	2.336	25.221
48 + 24	0.064	0.069	0.561	19.98	1.790	22.465
S.D.c	0.007	0.006	0.437	1.29	0.382	1.998

[a]Mean data from a four replicate composite for the two seasons. [b]TU/TS ratio = (total unsaturated fatty acids)/(total saturated fatty acids). [c]S.D. = standard deviation [37].

Table 18: Effect of N rate and foliar application of K and mepiquat chloride (MC) on the relative percentage of unsaturated fatty acids[a]

Treatments	Relative % of unsaturated fatty acids			TU/TS[b] ratio
	Oleic	Linoleic	Total	
N rate (kg·ha^{-1})				
95.2	21.59	53.65	75.24	3.069
142.8	22.99	54.08	77.06	3.397
S.D.[c]	1.35	1.14	2.28	0.403
K rate (g. ha-l)				
0	21.26	52.26	73.53	2.790

319	22.11	54.10	76.20	3.228
638	22.60	54.50	77.09	3.390
957	23.18	54.60	77.78	3.523
S.D[c]	1.37	0.	1.92	0.351
MC rate (g ha[-1])				
0	21.27	53.51	74.77	2.974
48 + 24	23.31	54.22	77.53	3.451
S.D.[c]	1.09	1.10	1.99	0.349

[a]Mean data from a four replicate composite for the two seasons. [b]TU/TS ratio = (total unsaturated fatty acids)/(total saturated fatty acids). [c]S.D. = standard deviation [37].

Seo et al. [105] found that, when sesame was given 0 to 160 kg N, oleic acid content was highest at the highest N rates and linoleic acid content was highest at the intermediate rates. Khan et al. [88] indicated that, oleic acid increased by increasing levels of N added to rapeseed-mustard. Kheir et al. [7], in flax, found that the higher N-rate increased the percentage of unsaturated fatty acids and decreased saturated fatty acids in the seed oil. Potassium applied at all concentrations resulted in a decrease in the total saturated fatty acids (capric, lauric, myristic, palmitic and stearic) compared with the untreated control (Table 17) [37]. Spraying plants with the high K concentration 957 g·K·ha^{-1} gave the lowest total saturated fatty acids oil, compared with the other two concentrations (638 and 957 g·K·ha^{-1}). The effect was significant for the two concentrations 638 and 957 g·K·ha^{-1} on capric, and palmitic, and for all different concentrations on lauric, myristic, stearic, and the total saturated fatty acids. Potassium applied at all rates increased the total unsaturated fatty acid (oleic and linoleic) and TU/TS ratio (by 1.84 - 4.48, and 15.70% - 26.27%, respectively), compared with untreated control (Table 18) [37]. Applied K at 957 g·ha^{-1} gave the highest increment, followed by 638 g·ha^{-1} concentration. The effect was significant for all different concentrations on linoleic, the total unsaturated fatty acid and TU/TS ratio [37]. Linoleic acid was the most abundant unsaturated fatty acid. The beneficial effect of applied K on TU and TU/TS ratio suggests that it might be due to the regulated effect of K which acts as an activator on many enzymic processes, where some of these enzymes may affect the seed oil content from these organic matters. Seo et al. [105] found that, when sesame was given 0 to180 kg K$_2$O, oleic acid content was the highest at the highest K rates and linoleic acid content was the highest at the intermediate rates. Salama [106] indicated that, K fertilizer applied to sunflower, favored fatty acid composition (high oleic acid content). Mekki et al. [84] stated that, foliar application with K on sunflower increased the oleic acid fatty acid. Froment et al. [85] found that, linoleic acid content was greatest in linseed oil in treatments receiving extra K. The application of MC resulted in a decrease in the total saturated fatty acids, the abundant saturated fatty acid palmitic, capric, myristic, and stearic, while it resulted in an increase in lauric saturated fatty acid, compared to the untreated control (Table 17) [37]. The effect was significant only on capric, palmitic, stearic and the total. The application of mepiquat chloride resulted in an increase in total unsaturated fatty acids (oleic and linoleic) and TU/TS ratio (by 3.69, and 16.69%, respectively), over the control (Table 18). The effect was significant only on the total unsaturated fatty acid, oleic and TU/TS ratio [37]. The stimulatory residual effects of the application mepiquat chloride on TU and TU/TS ratio was probably due to its favorable effects on fundamental metabolic reactions in plant tissues, and would have direct impact through utilization on growth processes, which are reflected in distinct changes in seed oil quality [37]. Some of these changes may affect the seed oil fatty acids composition, which may attribute to their encouraging effects on enzymes that catalyzed the biosynthesis of unsaturated fatty acids. Mekki and El-Kholy [107] investigated the response of rape oilseed to 0, 200 or 400 ppm mepiquat chloride and found that; palmitic acid was only decreased by using 400 ppm mepiquat chloride as compared with 200-ppm treatment or control plants. A low content of saturated fatty acids is desirable for edible purposes. Also, regarding oil quality, higher levels of linoleic acid and oleic acid are considered good for oil quality [108].

CONCLUSIONS

From the findings of this study, it seems rational to recommended application of N at a rate of 161 of kg·ha^{-1}, spraying of cotton plants with plant PGR, and application of Zn in comparison with the ordinary cultural practices adopted by Egyptian cotton producers, it is quite apparent that applications of such PGR, Zn, and increased N fertilization rates could bring about better impact on cottonseed yield, seed protein content, oil and protein yields, oil refractive index, unsaponifiable matter, and unsaturated fatty acids. On the other hand, there was a decrease in acid value and saponification value. The increase in seed yield and subsequent increase in oil and meal due to the application of PGR, Zn, and increased N fertilization were sufficient to cover the cost of using those chemicals and further attain an economical profit [35].

It can be concluded that addition of P at 74 kg·ha^{-1}, and foliar application of Zn and Ca at different concentrations (especially Ca concentration of 60 ppm) beneficially affected cottonseed yield, seed index, seed oil content, oil and protein yields ha^{-1}, seed oil unsaponifiable matter, and total unsaturated fatty acids (oleic and linoleic) [36].

The addition of K at 47.4 kg·ha^{-1}, spraying cotton plants with Zn twice (at 57.6 g·ha^{-1}), and also with P twice (especially the P concentration of 1728 g·ha^{-1}) along with the soil fertilization used P at sowing time have been proven beneficial to the quality and yield of cotton plants. These combinations appeared to be the most effective treatments, affecting not only the quantity but also the quality of oil, and to obtain higher oil and protein yields and a better fatty acid profile in the oil of cotton. In comparison with the ordinary cultural practices adopted by Egyptian cotton producers, it is apparent that the applications of such treatments could produce an improvement in cottonseed yield, seed protein content, oil and protein yields, oil refractive index, unsaponifiable matter, iodine value, unsaturated fatty acids and a decrease in oil acid value and saponification value. The increase in seed yield and subsequent increase in oil and meal due to the addition of K, spraying cotton plants with Zn and of P are believed to be sufficient enough to cover the cost of using those chemicals and obtain an economic profit at the same time [1].

Application of N at the rate of 143 kg·ha^{-1} and two applications of both K (foliar; at the rate of 957 g·K·ha^{-1}) and mepiquat chloride (at a rate of 48 + 24 g a.i. ha^{-1}, respectively) have the most beneficial effects among the treatments examined, affecting not only the seed quantity (to obtain higher oil and protein yields ha^{-1}) but also the oil seed quality (as indicated by better fatty acid profile in the oil of cotton) in comparison with the usual cultural practices adopted by Egyptian cotton procedures [37].

REFERENCES

1. Sawan, Z.M., Hafez, S.A., Basyony, A.E. and Alkassas, A.R. (2007) Cottonseed: Protein, oil yields, and oil properties as influenced by potassium fertilization and foliar application of zinc and phosphorus. Grasas Y. Aceites, 58, 40-48.
2. Reddy, A.R., Reddy, K.R., Padjung, R. and Hodges, H.F. (1996) Nitrogen nutrition and photosynthesis in leaves of Pima cotton. Journal of Plant Nutrition, 19, 755-770. http://dx.doi.org/10.1080/01904169609365158
3. Rinehardt, J.M., Edmisten, K.L., Wells, R. and Faircloth, J.C. (2004) Response of ultra-narrow and conventional spaced cotton to variable nitrogen rates. Journal of Plant Nutrition, 27, 743-755. http://dx.doi.org/10.1081/PLN-120030379
4. Ansari, M.S. and Mahey, R.K. (2003) Growth and yield of cotton species as affected by sowing dates and nitrogen levels. Punjab Agricultural University, Journal of Research, 40, 8-11.
5. Patil, B.N., Lakkineni, K.C. and Bhargava, S.C. (1996) Seed yield and yield contributing characters as influenced by N supply in rapeseed-mustared. Journal of Agronomy and Crop Science, 177, 197-205. http://dx.doi.org/10.1111/j.1439-037X.1996.tb00237.x
6. Frink, C.R., Waggoner, P.E. and Ausubel, J.H. (1999) Nitrogen fertilizer: Retrospect and prospect. Proceedings of the National Academy of Sciences of the United States of America, 96, 1175-1180. http://dx.doi.org/10.1073/pnas.96.4.1175
7. Kheir, N.F., Harb, E.Z., Moursi, H.A. and El-Gayar, S.H. (1991) Effect of salinity and fertilization on flax plants (Linum usitatissimum L.). II. Chemical composition. Bulletin of Faculty of Agriculture-University of Cairo, 42, 57-70.
8. Taiz, L. and Zeiger, E. (1991) Plant physiology: Mineral nutrition. The Benjamin Cummings Publishing Co., Inc. Redwood City, 100-119.
9. Rodriguez, D., Zubillaga, M.M., Ploschuck, E., Keltjens, W., Goudriaan, J. and Lavado, R. (1998) Leaf area expansion and assimilate prediction in sunflower growing under low

phosphorus conditions. Plant and Soil, 202, 133- 147.http://dx.doi.org/10.1023/A:1004348702697

10. Sasthri, G., Thiagarajan, C.P., Srimathi, P., Malarkodi, K. and Venkatasalam, E.P. (2001) Foliar application of nutrient on the seed yield and quality characters of nonaged and aged seeds of cotton cv. MCU5. Madras Agricultural Journal, 87, 202-206.

11. Stewart, W.M., Reiter, J.S. and Krieg, D.R. (2005) Cotton response to multiple application of phosphorus fertilizer. Better Crops with Plant Food, 89, 18-20.

12. Singh, V., Pallaghy, C.K. and Singh, D. (2006) Phosphorus nutrition and tolerance of cotton to water stress: I. Seed cotton yield and leaf morphology. Field Crops Research, 96, 191-198. http://dx.doi.org/10.1016/j.fcr.2005.06.009

13. Ibrahim, M.E., Bekheta, M.A., El-Moursi, A. and Gaafar, N.A. (2009) Effect of arginine, prohexadione-Ca, some macro and micro-nutrients on growth, yield and fiber quality of cotton plants. World Journal of Agricultural Sciences, 5, 863-870.

14. Gebaly Sanaa, G. and El-Gabiery, A.E. (2012) Response of cotton Giza 86 to foliar application of phosphorus and mepiquat chloride under fertile soil condition. Journal of Agricultural Research, 90, 191-205.

15. Sangakkara, U.R., Frehner, M. and Nösberger, J. (2000) Effect of soil moisture and potassium fertilizer on shoot water potential, photosynthesis and partitioning of carbon in mungbean and cowpea. Journal of Agronomy and Crop Science, 185, 201-207. http://dx.doi.org/10.1046/j.1439-037x.2000.00422.x

16. Colomb, B., Bouniols, A. and Delpech, C. (1995) Effect of various phosphorus availabilities on radiation-use efficiency in sunflower biomass until anthesis. Journal of Plant Nutrition, 18, 1649-1658. http://dx.doi.org/10.1080/01904169509365010

17. Gormus, O. (2002) Effects of rate and time of potassium application on cotton yield and quality in Turkey. Journal of Agronomy and Crop Science, 188, 382-388.http://dx.doi.org/10.1046/j.1439-037X.2002.00583.x

18. Aneela, S., Muhammad, A. and Akhtar, M.E. (2003) Effect of potash on boll characteristics and seed cotton yield in newly

developed highly resistant cotton varieties. Pakistan Journal of Biological Sciences, 6, 813-815.

19. Pervez, H., Ashraf, M. and Makhdum, M.I. (2004) Influence of potassium rates and sources on seed cotton yield and yield components of some elite cotton cultivars. Journal of Plant Nutrition, 27, 1295-1317. http://dx.doi.org/10.1081/PLN-120038549

20. Pettigrew, W.T., Meredith Jr., W.R. and Young, L.D. (2005) Potassium fertilization effects on cotton lint yield, yield components, and reniform nematode populations. Agronomy Journal, 97, 1245-1251. http://dx.doi.org/10.2134/agronj2004.0321

21. Sharma, S.K. and Sundar, S. (2007) Yield, yield attributes and quality of cotton as influenced by foliar application of potassium. Journal of Cotton Research and Development, 21, 51-54.

22. Welch, R.M. (1995) Micronutrient nutrition of plants. Critical Reviews in Plant Sciences, 14, 49-82.

23. Oosterhuis, D., Hake, K. and Burmester, C. (1991) Leaf feeding insects and mites. Cotton council of America. Cotton Physiology Today, 2, 1-7.

24. Rathinavel, K., Dharmalingam, C. and Paneersel vam, S. (2000) Effect of micronutrient on the productivity and quality of cotton seed cv. TCB 209 (Gossypium barbadense L.). Madras Agricultural Journal, 86, 313-316.

25. Li, L.L., Ma, Z.B., Wang, W.L. and Tai, G.Q. (2004) Effect of spraying nitrogen and zinc at seedling stage on some physiological characteristics and yield of summer cotton. Journal of Henan Agricultural University, 38, 33-35.

26. Rensing, L. and Cornelius, G. (1980) Biological membranes as components of oscillating systems. Biologische Rundschau, 18, 197-209.

27. Ma, L.G. and Sun, D.Y. (1997) The involvement of calcium in the light signal transduction chain for phototropism in sunflower seedling. Biologia Plantarum, 39, 569-574.

28. Ochiai, E.L. (1977) Bioinorganic chemistry. Allyn & Bacon, Boston, 515 p.

29. Zhao, D.L. and Oosterhuis, D.M. (2000) Pix plus and mepiquat chloride effects on physiology, growth, and yield of field-grown cotton. Journal of Plant Growth Regulation, 19, 415-422.
30. Wang, Z.L., Yin, Y.P. and Sun, X.Z. (1995) The effect of DPC (N, N-dimethyl piperidinium chloride) on the $^{14}CO_2$- assimilation and partitioning of ^{14}C assimilates within the cotton plants interplanted in a wheat stand. Photosynthetica, 31, 197-202.
31. Kumar, K.A.K., Patil, B.C. and Chetti, M.B. (2004) Effect of plant growth regulators on biophysical, biochemical parameters and yield of hybrid cotton. Karnataka Journal of Agricultural Sciences, 16, 591-594.
32. Palomo Gil, A. and Chávez González, J.F. (1997) Response of the early cotton cultivar CIAN 95 to nitrogen fertilizer application. ITEA. Producción Vegetal, 93, 126- 132.
33. Sarwar Cheema, M., Akhtar, M. and Nasarullah, M. (2009) Effect of foliar application of mepiquat chloride under varying nitrogen levels on seed cotton yield and yield components. Journal of Agricultural Research, 47, 381-388.
34. Patil, D.B., Naphade, K.T., Wankhade, S.G., Wanjari, S.S. and Potdukhe, N.R. (1997) Effect of nitrogen and phosphate levels on seed protein and carbohydrate content of cotton cultivars. Indian Journal of Agricultural Research, 31, 133-135.
35. Sawan, Z.M., Hafez, S.A. and Basyony, A.E. (2001) Effect of nitrogen and zinc fertilization and plant growth retardants on cottonseed, protein, oil yields, and oil properties. Journal of the American Oil Chemists' Society, 78, 1087-1092.
36. Sawan, Z.M., Hafez, S.A. and Basyony, A.E. (2001) Effect of phosphorus fertilization and foliar application of chelated zinc and calcium on seed, protein and oil yields and oil properties of cotton. Journal of Agricultural Science, 136, 191-198.
37. Sawan, Z.M., Hafez, S.A., Basyony, A.E. and Alkassas, A.R. (2007) Nitrogen, potassium and plant growth retardant effects on oil content and quality of cotton seed. Grasas Y. Aceites, 58, 243-251.
38. Sawan, Z.M., Fahmy, A.H. and Yousef, S.E. (2009) Direct and residual effects of nitrogen fertilization, foliar application of potassium and plant growth retardant on Egyptian cotton growth,

seed yield, seed viability and seedling vigor. Acta Ecologica Sinica, 29, 116-123. http://dx.doi.org/10.1016/j.chnaes.2009.05.008
39. Association of Official Analytical Chemists (1985) Official methods of analysis. 14th Edition, Association of Analytical Communities (AOAC), Arlington.
40. Kates, M. (1972) Laboratory techniques in biochemistry and molecular biology. In: Work, T.S. and Work, E., Eds., North-Holland Publishing, Amsterdam.
41. American Oil Chemists' Society (1985) Official methods and recommended practices of the american oil chemists' society. In: Walker, R.O., Ed., 3rd Edition, American Oil Chemists' Society, Champaign.
42. Vogel, A.I. (1975) A textbook of practical organic chemistry. 3rd Edition, English Language Book Society and Longman Group, Harlow.
43. Ashoub, A.H., Basyony, A.E. and Ebad, F.A. (1989) Effect of plant population and nitrogen levels on rapeseed oil quality and quantity. Annals of Agricultural Science, Moshtohor, 27, 761-770.
44. Snedecor, G.W. and Cochran, W.G. (1980) Statistical methods. 7th Edition, Iowa State University Press, Ames.
45. Abdel-Malak, K.I., Radwan, F.E. and Baslious, S.I. (1997) Effect of row width, hill spacing and nitrogen levels on seed cotton yield of Giza 83 cotton cultivar. Egyptian Journal of Agricultural Research, 75, 743-752.
46. Saleem, M.F., Shakeel, A., Bilal, M.F., Shahid, M.Q. and Anjum, S.A. (2010) Effect of different phosphorus levels on earliness and yield of cotton cultivars. Soil & Environment, 29, 128-135.
47. Hamed, F.S., Abo El-Hamd, A.S., Ibrahim, M.M. and ElSayed, A.E.M. (2012) Effect of some cultural practices on growth, flowering, earliness characters and yield of cotton plant variety Giza 90 (Gossypium barbadense L.). Egyptian Journal of Agricultural Research, 90, 1649-1673.
48. Gardner, F.P. (1988) Growth and partitioning in peanut as influenced by gibberellic acid and daminozide. Agronomy Journal, 80, 159-163. http://dx.doi.org/10.2134/agronj1988.00021962008000020004x

49. Nepomuceno, A.L., Oosterhuis, D.M. and Steger, A. (1997) Duration of activity of the plant growth regulators PGR-IV and mepiquat chloride. Special Reports-Agricultural Experiment Station, Division of Agriculture, University of Arkansas, Arkansas, No. 183, 136-139.
50. Abdel-Al, M.H. (1998) Response of Giza 85 cotton cultivar to the growth regulators Pix and Atonic. Egyptian Journal of Agricultural Research, 76, 1173-1181.
51. Pípolo, A.E., Athayde, M.L.F., Pípolo, V.C. and Parducci, S. (1993) Comparison of different rates of chlorocholine chloride applied to herbaceous cotton. Pesquisa Agropecuária Brasileira, 28, 915-923.
52. Sawan, Z.M., Basyony, A.E., McCuistion, W.L. and ElFarra, A.A. (1993) Effect of plant population densities and application of growth retardants on cottonseed yield and quality. Journal of the American Oil Chemists' Society, 70, 313-317.
53. Zeng, Q.F. (1996) Researchers on the effect of zinc applied to calcareous soil in cotton field. China Cottons, 23, 21.
54. Carvalho, L.H., Chiavegato, E.J., Cia, E., Kndo, J.I., Sabino, J.C., Pettinelli Junior, A., Bortoletto, N. and Gallo, P.B. (1994) Plant growth regulators and pruning in the cotton crop. Bragantia, 53, 247-254. http://dx.doi.org/10.1590/S0006-87051994000200014
55. Pandrangi, R.B., Wankhade, S.G. and Kedar, G.S. (1992) Response of cotton (Gossypium hirsutum L.) to N and P grown under rainfed conditions. Crop Research, 5, 54-58.
56. Sawan, Z.M., Sakr, R.A., Ahmed, F.A. and Abd-Al-Samad, A.M. (1991) Effect of 1,1-dimethyl piperidinium chloride (Pix) on the seed, protein, oil and fatty acids of Egyptian cotton. Journal of Agronomy and Crop Science, 166, 157- 161. http://dx.doi.org/10.1111/j.1439-037X.1991.tb00899.x
57. Sawan, Z.M., El-Farra, A.A. and Mohamed, Z.A. (1988) Effect of nitrogen fertilization, foliar application of calcium and some micro-elements on cottonseed, protein and oil yields and oil properties of Egyptian cotton. Annali di Botanica, 46, 167-174.
58. Prabhuraj, D.K., Badiger, M.K. and Manure, G.R. (1993) Growth and yield of sunflower (Helianthus annuus) as influenced by levels

of phosphorus, sulphur and zinc. Indian Journal of Agronomy, 38, 427-430.
59. Bybordi, A. and Mamedov, G. (2010) Evaluation of application methods efficiency of zinc and iron for canola (Brassica napus L.). Notula Scientia Biologicae, 2, 94- 103.
60. Sugiyama, T., Mizuno, M. and Hayashi, M. (1984) Partitioning of nitrogen among Ribulose-1, 5-bisphosphate carboxylase/ oxygenase, phosphoenolpyruvat carboxylase, and pyruvate orthophosphate dikinase as related to biomass productivity in maize seedlings. Plant Physiology, 75, 665-669. http://dx.doi.org/10.1104/pp.75.3.665
61. Hedin, P.A., McCarthy, J.C. and Jenkins, J.N. (1988) Effect of CCC and PIX related bioregulators on gossypol, protein, yield and seed properties of cotton. Journal of the Mississippi Academy of Sciences, 33, 49-57.
62. Kar, C., Barua, B. and Gupta, K. (1989) Response of the safflower plant (Carthamus tinctorius L. cv. JLA 900) towards plant growth retardants dikegulac sodium, CCC and SADH. Indian Journal of Plant Physiology, 23, 144- 147.
63. Wang, H.Y. and Chen, Y. (1984) A study with ^{32}P on the effect of growth regulators on the distribution of nutrients with cotton plants. China Cottons, 4, 29-30.
64. Kler, D.S., Raj, D. and Dhillon, G.S. (1991) Chemical regulated growth, development and light penetration in American cotton (Gossypium hirsutum L.). Environment and Ecology, 9, 584-588.
65. Osman, R.O. and Abu-Lila, B.H. (1985) Studies on the effect of gibberellic acid and Cycocel on flax plants (Linum usitatissimum L.), seed oil content and oil composition. Zeitschrift für Acker- und Pflanzenbau, 155, 82-88.
66. Guinn, G. (1984) Boll abscission in cotton. In: Gupta, U.S., Ed., Crop Physiology: Advancing Frontiers, Mohan Primlani for Oxford & IBH Publishing Co., New Delhi, 177-225.
67. Saleem, M.F., Shakeel, A., Bilal, M.F., Shahid, M.Q. and Anjum, S.A. (2010) Effect of different phosphorus levels on earliness and yield of cotton cultivars. Soil & Environment, 29, 128-135.
68. Glass, A.D.M. (1989) Plant nutrition. An introduction to current concepts. Jones and Bartlett Publishers, Boston/ Portola Valley.

69. Shui, J.G. and Meng, S.F. (1990) Effects of lime application on cotton yield in red soil fields. China Cottons, 1, 26-29.
70. Wright, S.D., Munk, D., Munier, D., Vargas, R., Weir, B., Roberts, B. and Jimenez Jr., M. (1995) Effect of aminofol/ boll-set plus calcium zinc on California cotton. Proceedings Beltwide Cotton Conferences, San Antonio, TX, USA, 4-7 January 1995, Volume 2, National Cotton Council, Memphis.
71. Kosheleva, L.L., Bakhnova, K.V., Semenova, T.A. and Mil'Kevich, Z.A. (1984) Effect of phosphorus nutrition on metabolism of young fiber flax plants in relation to assimilate distribution in them. Referativnyi Zhurnal, 6, 533.
72. El-Debaby, A.S., Hammam, G.Y. and Nagib, M.A. (1995) Effect of planting date, N and P application levels on seed index, lint percentage and technological characters of Giza 80 cotton cultivar. Annals of Agricultural Science, Moshtohor, 33, 455-464.
73. Bora, P.C. (1997) Effect of gypsum and lime on performance of Brassica varieties under rainfed condition. Indian Journal of Agronomy, 42, 155-158.
74. Davidson, F.M. and Long, C.M. (1958) The structure of the naturally occurring phosphoglycerides. 4. Action of cabbage leaf phospholipase. Biochemical Journal, 69, 458.
75. Shchitaeva, V.A. (1984) Effect of zinc on metabolic activity of the root system of fine-fibered cotton. Izvestiya Akademii Nauk Turkmenskoi SSR, Biologicheskaya, 4, 8-13.
76. Gushevilov, Zh. and Palaveeva, Ts. (1991) Effect of longterm systematic fertilizer application on sunflower yield and quality. Pochvoznanie i Agrokhimiya, 26, 20-26.
77. Gebaly Sanaa, G. (2012) Physiological effects of potassium forms and methods of application on cotton variety Giza 80. Egyptian Journal of Agricultural Research, 90, 1633-1646.
78. Russell, E.W. (1973) Soil condition and plant growth. The English Language Book Society and Longman, London, 448 p.
79. Ghourab, M.H.H., Wassel, O.M.M. and Raya, N.A.A. (2000) Response of cotton plants to foliar application of (Pottasin-P)™ under two levels of nitrogen fertilizer. Egyptian Journal of Agricultural Research, 78, 781-793.
80. Wiatrak, P.J., Wright, D.L. and Marois, J.J. (2006) Development and yields of cotton under two tillage systems and nitrogen

application following white lupine grain crop. Journal of Cotton Science, 10, 1-8.
81. Cakmak, I. (2000) Possible roles of zinc in protecting plant cells from damage by reactive oxygen species. New Phytologist, 146, 185-205. http://dx.doi.org/10.1046/j.1469-8137.2000.00630.x
82. Bednarz, C.W. and Oosterhuis, D.M. (1999) Physiological changes associated with potassium deficiency in cotton. Journal of Plant Nutrition, 22, 303-313.http://dx.doi.org/10.1080/01904169909365628
83. Marschner, H. (1995) Mineral nutrition of higher plants. 2nd Edition, Academic Press, London.
84. Mekki, B.B., El-Kholy, M.A. and Mohamed, E.M. (1999) Yield, oil and fatty acids content as affected by water deficit and potassium fertilization in two sunflower cultivars. Egyptian Journal of Agronomy, 21, 67-85.
85. Froment, M.A., Turley, D. and Collings, L.V. (2000) Effect of nutrition on growth and oil quality in linseed. Tests of Agrochemicals and Cultivars, 21, 29-30.
86. Klug, A. and Rhodes, D. (1987) "Zinc fingers": A novel protein motif for nucleic acid recognition. Trends in Biochemical Sciences, 12, 464-469.http://dx.doi.org/10.1016/0968-0004(87)90231-3
87. Romheld, V. and Marschner, H. (1991) Micronutrients in agriculture. 2nd Edition, Soil Science Society of America Book Series, no 4, Soil Science Society of America, Inc., Madison, Wisconsin, 297-328.
88. Khan, N.A., Ansari, H.R. and Samiullah (1997) Effect of gibberellic acid spray and Basal nitrogen and phosphorus on productivity and fatty acid composition of rapeseedmustard. Journal of Agronomy and Crop Science, 179, 29- 33. http://dx.doi.org/10.1111/j.1439-037X.1997.tb01144.x
89. Legé, K.E., Cothren, J.T. and Morgan, P.W. (1997) Nitrogen fertility and leaf age effects on ethylene production of cotton in a controlled environment. Plant Growth Regulation, 22, 23-28.
90. McConnell, J.S. and Mozaffari, M. (2004) Yield, petiole nitrate, and node development responses of cotton to early season nitrogen fertilization. Journal of Plant Nutrition, 27, 1183-1197. http://dx.doi.org/10.1081/PLN-120038543

91. Pettigrew, W.T. (1999) Potassium deficiency increases specific leaf weights of leaf glucose levels in field-grown cotton. Agronomy Journal, 91, 962-968.http://dx.doi.org/10.2134/agronj1999.916962x

92. Cakmak, I., Hengeler, C. and Marschner, H. (1994) Partitioning of shoot and root dry matter and carbohydrates in bean plants suffering from phosphorus, potassium and magnesium deficiency. Journal of Experimental Botany, 45, 1245-1250.http://dx.doi.org/10.1093/jxb/45.9.1245

93. Mullins, G.L., Schwab, G.J. and Burmester, C.H. (1999) Cotton response to surface applications of potassium fertilizer: A 10-year summary. Journal of Production Agriculture, 12, 434-440. http://dx.doi.org/10.2134/jpa1999.0434

94. Prakash, R., Prasad, M. and Pachauri, D.K. (2001) Effect of nitrogen, chlormequat chloride and FYM on growth yield and quality of cotton (Gossypium hirsutum L.). Annals of Agricultural Research, 22, 107-110.

95. Mekki, B.B. (1999) Effect of mepiquat chloride on growth, yield and fiber properties of some Egyptian cotton cultivars. Arab University Journal of Agricultural Science, 7, 455-466.

96. Palomo Gil, A., Godoy Avila, S. and Chávez González, J.F. (1999) Reductions in nitrogen fertilizers use with new cotton cultivars: Yield, yield components and fiber quality. Agrociencia, 33, 451-455.

97. Ali, S.A. and El-Sayed, A.E. (2001) Effect of sowing dates and nitrogen levels on growth, earliness and yield of Egyptian cotton cultivar Giza 88. Egyptian Journal of Agricultural Research, 79, 221-232.

98. Lamas, F.M. (2001) Comparative study of mepiquat chloride and chlormequat chloride application in cotton. Pesquisa Agropecuária Brasileira, 36, 265-272.http://dx.doi.org/10.1590/S0100-204X2001000200008

99. Zubillaga, M.M., Aristi, J.P. and Lavado, R.S. (2002) Effect of phosphorus and nitrogen fertilization on sunflower (Helianthus annus L) nitrogen uptake and yield. Journal of Agronomy and Crop Science, 188, 267-274. http://dx.doi.org/10.1046/j.1439-037X.2002.00570.x

100. Madraimov, I. (1984) Potassium fertilizers and oil content of cotton seeds. Khlopkovodstvo, 6, 11-12.

101. Fan, S.L., Xu, Y.Z. and Zhang, C.J. (1999) Effects of nitrogen, phosphorus and potassium on the development of cotton bolls in summer. Acta Gossypii Sinica, 11, 24-30.

102. Stitt, M. (1999) Nitrate regulation of metabolism and growth. Current Opinion in Plant Biology, 2, 178-186. http://dx.doi.org/10.1016/S1369-5266(99)80033-8

103. Narang, R.S., Mahal, S.S. and Gill, M.S. (1993) Effect of phosphorus and sulphur on growth and yield of toria (Brassica campestris subsp. oleifera var toria). Indian Journal of Agronomy, 38, 593-597.

104. Holmes, M.R.J. and Bennett, D. (1979) Effect of nitrogen fertilizer on the fatty acid composition of oil from low erucic acid rape varieties. Journal of the Science of Food and Agriculture, 30, 264-266. http://dx.doi.org/10.1002/jsfa.2740300309

105. Seo, G.S., Jo, J.S. and Choi C.Y. (1986) The effect of fertilization level on the growth and oil quality in sesame (Sesamum indicum L.). Korean Journal of Crop Science, 31, 24-29.

106. Salama, A.M. (1987) Yield and oil quality of sunflowers as affected by fertilizers and growth regulators. Növénytermelés, 36, 191-202.

107. Mekki, B.B. and El-Kholy, M.A. (1999) Response of yield, oil and fatty acid contents in some oil seed rape varieties to mepiquat chloride. Bulletin of the National Research Center, 24, 287-299.

108. Downey, R.K. and Rimmer, S.R. (1993) Agronomic improvement in oil seed Brassicas. Advances in Agronomy, 50, 1-66. http://dx.doi.org/101016/S0065-2113(08)60831-7

Chapter 10

Process Adaption and Modifications of a Nutrient Removing Wastewater Treatment Plant in Sri Lanka Operated at Low Loading Conditions

Johanna Berg[1] and Stig Morling[2]

[1]Purac AB, Emdalavägen, Lund, Sweden
[2]SWECO Environment, Stockholm, Sweden

ABSTRACT

The Sri Lankan national water authority, that is The National Water Supply and Drainage Board (NWS&DB) has taken a new wastewater treatment plant into operation at Ja Ela, North of Colombo. The plant has been in operation since September 2011. In April 2012, it was concluded how a test of the aeration efficiency and a performance test should be carried out. The tests have been based on the actual loading of the plant and the analysis results from the daily process control. The evaluation of the aeration efficiency is not reported in this paper. The paper presents the overall performance of the water treatment part of the plant during start-up conditions, from fall 2011 through the first five months of 2012. The results from the operation are found in Table 1. An important circumstance at the plant is the current very low loading in comparison with the design load. This fact has resulted in an introduction of an intermittent mode of the aeration (nitrification) reactor. Based on operation figures, during more than a month (May 2012), it has been possible to give a realistic assessment of the overall performance. The most striking results are summarized as follows: 1) The intermittent operation has enabled an energy efficient operation of the plant. By the introduction of the intermittent aeration, the energy consumption has been reduced by around 75%, compared with the continuous operation mode; 2) The plant performance during the intermittent operation has been improved with respect to virtually all important pollution variables. The most striking improvement is the discharge total P level, reflecting that a substantial enhanced biological phosphorus removal takes. The typical discharge levels found during May 2012, were compared with the earlier obtained values. It is important to underline that the loading on the plant has slightly increased during May as compared with the previous operation period.

BACKGROUND

Planning and construction of wastewater treatment plants is normally a very demanding task that includes a large number of assumptions, considerations and technical knowledge. Within the field of municipal wastewater planning, some distinct points should be considered. These points are not necessarily identical with what is found for industrial wastewater planning. Some major conditions that rule the planning, design and construction of a municipal plant, may be summarized as follows:

- An identified design and operation period is normally defined at an early stage of the planning phase. A very often used operation period is found to be between 15 and 20 years;

- There is often a more or less strong political demand for a "sustainable" wastewater treatment. Even though the expression is somewhat unclear it is more or less imperative to include aspects on sustainability in the planning. Of course you may argue that any professional planning and design should by necessity be based on a technical sustainability;
- The operation period will normally include a step by step increase of pollution loading of the plant during the operation time. Therefore, most municipal plants are designed for a specific loading occurring at the end of the design period; say 15 to 20 years away. This design capacity is defined to allow for a growth of the community within the catchment area. The opposite mode (to design the plant for the current pollution load only) would create political, technical and operational problems!
- Another important circumstance is the fact that the community has limited financial resources. Thus it has to relate both to the sizing of the plant as well as to the plant configuration;
- In order to enlighten the situation another circumstance would be addressed. The given effluent standards for a municipal WWTP are normally valid for a limited time, defined in the legal permit. We may call this for the "environmental lifetime" for the performed investment. Looking back in the latest century it is easy to find a more or less typical time frame for the permit. The example from Sweden seems to sustain the assumption of a 20 year time for an update and sharpening of the effluent standards.

Table 1: Summary of the discharge levels from Ja-Ela WWTP in May 2012 compared with earlier typical effluent values

	January-April 2012	May 2012
BOD_5		
Nos of observ.	14	8
Mean value. mg/l	5.5	2

Consent value, mg/l	<25	<25
COD		
Nos of observ.	26	15
Mean value. mg/1	48	35.9
Consent value. mg/	<120	<120
Total N		
Nos of observ.	23	15
Mean value. mg/1	10.4	7.2
Consent value. mg/l	<30	<30
Total P		
Nos of observ.	11	15
Mean value. mg/1	4.6	1.2
Consent value, mg/l	<5	<5

These conditions together form some of the pre-requisites for a municipal WWTP. One of the most common and sometimes surprising situations is anyhow that a plant will operate substantially below the design load conditions. If not addressed properly, this may result in very poor operation economics—excessive operation costs per amount of pollution treated. In some cases it may also lead to inferior performance. One possible cause may be a nitrification that "burns out" all alkalinity and thus more or less inhibits the entire process, due to a pH drop.

Historically a large number of plant operators have experienced this situation. An efficient way to mitigate such a situation has been to change the operation mode, either by closing part of the plant, or to introduce an intermittent operation. A probably well-known example of converting the operation of a continuously working plant into intermittent operation was the development of the BioDenitro system in Denmark [1]. Further basic and applied knowledge derived from the SBR-technology development can be obtained on intermittent operation strategies, [2-5]. The demands for an improved control of the activated sludge system by means on on-line instruments, and of course most of all a deep understanding of the process kinetics has been addressed recently [6]. The potential to improve the process performance and at the same time save aeration energy is obvious even for plants operated at almost design conditions.

The Ja-Ela/Ekala WWTP is part of a larger scheme covering both the collection systems and a second WWTP, south of Colombo. Financing has been secured by support from Sida (Swedish International Development Agency), the basic design work and engineering support has been performed by SWECO, Sweden. The plant design and construction has been performed by Purac AB, Sweden.

The situation with a low load situation during the startup year at the Ja-Ela/Ekala WWTP may be seen as a very typical example of an "under-loaded" plant. So this plant presents a rather typical example of the initial operating situation. The challenge in this case has been to address and mitigate the problem in a cultural environment with limited experience of running advanced wastewater treatment facilities.

This paper describes the situation with the initially encountered problems with the under-loading, as well as the actions taken to mitigate the identified problems. The outcomes of the alteration of the plant are accordingly reported, thanks to a very frequent sampling and analysis of the wastewater and sludge.

MATERIAL AND METHODS

Design Figures and Permissible Discharge Levels for the Ja-Ela/Ekala WWTP

The following Table 2 shows the design loads for the Ja-Ela/Ekala WWTP, along with the anticipated load developments. The following table illustrates the presumed development of the flows and loads during the design period. The plant configuration is shown in the flow sheet found in Figure 1.

Sampling and Analysis

As the accredited laboratory was not been able to present the analysis results within an acceptable timeframe it was concluded to use the onsite resources. Thus, the presented results are based on a site laboratory analysis, using the Hach-Lange analysis kits. The methods are defined as follows:

For BOD test: LCK 555;

For COD test on influent: LCK 514;

For COD test on effluent: LCK 314;

For Total N test on influent: LCK 338;

For Total N test on effluent: LCK 238;

For NH_4-N tests: LCK 303;

For Nitrate test on effluent: LCK 340;

For Total P test on influent: LCK 338;

For PO_4-P tests: LCK 348.

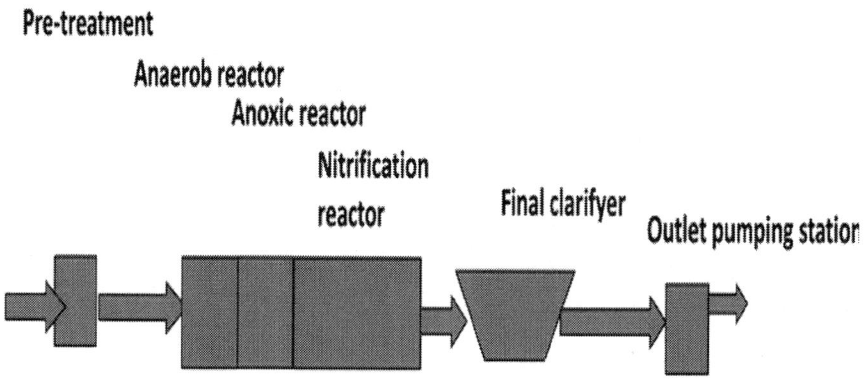

Figure 1: Simplified flow sheet over Ja-Ela/Ekala WWTP.

Table 2: Main design figures and ruling consent levels for Ja-Ela/Ekala WWTP

Stage	First Stage	Final Stage	
Inlet flows			
Total design flow per hour, m³/h	410	819	
Total daily design flow, m³/d	7250	14.500	
Pollution loads. kg/d			
BOD5	2191	4382	
COD	6135	12.269	
SS	2421	4841	
Total N	441	881	
Total P	104	208	
Ratios, kg/kg			
COD/BOD$_5$	2.80	2.80	
SS/BOD$_5$	1.10	1.10	
BOD$_5$/N	4.97	4.97	
BOD$_5$/P	21.1	21.1	
Concentrations, mg/l			Consent levels
BOD$_5$	302	302	<25
COD	846	846	<120
SS	334	334	<35
Total N	61	61	<30
NH$_4$-N	45	45	<10
Total P	14	14	<5
Max. Temperature, C	<40	<40	<40

For tests on suspended solids (SS), volatile suspended solids, of dry solids (DS) or total solids (TS) and volatile solids (VS) the contractor has provided standard instructions.

For inlet and discharge sampling automatic samplers are installed and have been used for 24 hour composite sampling, performed with an ASP station 2000 Peristaltic. The free oxygen level in the nitrification tank is measured on line with an Endress Hauser Liquisys com 223/253 meter. In a similar manner the SS-concentration is measured on line in the biological reactors by using an Endress Hauser Liquisys cum 223/253 meter. The wastewater flow is measured and recorded on a daily basis. Two different flow measurement points are defined, 1) at the inlet to the biological treatment, where an overflow arrangement is installed; 2) At the discharge point from the plant, after the effluent pumping station, where a magnetic flow meter is installed.

PERFORMANCE AND RESULTS

Flows into the Plant

The plant has been in operation since September 2011. In principle the flow is recorded on a daily basis. The two measurement points at the inlet and the outlet differ to a certain extent. Thus, the following presentation and analysis is based on the flow measurements at the discharge point, as the magnetic flow meter is deemed to provide far more correct figures than the overflow arrangement provides. The number of flow observations is adequate to allow a presentation of statistics. This is found in Table 3.

Pollution Loads into the Plant

The amounts of analysis results on incoming loads are far more limited than the recording of daily flows. In the following the main pollutants are presented statistically, though the recording is limited to year 2012.

Organic Pollution

The organic loading of the plant is measured mainly as COD, although the BOD_5 also is presented with a few observations. Shown in Table 4 are the concentrations and loads for BOD_5 and COD during the period January 2012 through to mid-April 2012. As a comparison the corresponding figures from May 2012 are shown in the same table. Comments to Table 4: It is important to observe that the number of observations with respect to BOD is very limited. Thus it is preferred to use the median value in the following comments. The design load expressed as BOD is 2191 kg/d. Thus the median loading is only 10.5% of the design level.

Nutrient Loading

In a similar manner the nutrients, nitrogen and phosphorrus are recorded during the period. Also for these variables only the 2012 year figures are presented. In order to make the comparison somewhat easier the two periods of January-April 2012 and May 2012 are shown side by side. In Table 5 is found the nitrogen values in incoming wastewater. Comments to Table 5: The number of observations is in both cases many enough to be used from a statistical point of view. As the mean and median values do not deviate substantially we will use the mean values for further discussions. The actual mean value is 13.1% of the design load. For the incoming phosphorus values recorded concentrations and calculated loads are presented in Table 6. Comments to Table 6: The number of observations with respect to phosphorus is limited. The mean and median values are very close, thus we use the mean value for further comments. The actual mean value is 8.7% of the design load.

Table 3: Summary of wastewater flows leaving the Ja-Ela WWTP from September 2011 through to April 2012, and for May 2012

Period	September 2011-April 2012	May 2012
Nos of observ.	175	33
Design flow, m³/d	7250	7250
Max value, m³/d	2668	1487
Mean, m³/d	903	991
Median, m³/d	906	1,013
Min value, m³/d	16	561
Standard deviation, m³/d	338	189
Standard error, m³/d	25.6	32.9

Table 4: Summary of organic loads (BOD5) and COD into Ja-Ela WWTP January 2012 through to April 2012, and for May 2012

Period	January- April 2012		May 2012	
BOD_5	Conc.	Loads	Conc.	Loads
Value	mg/l	kg BOD_3/d	mg/l	kg BOD_3/d
Nos of observ.	8	5	7	7
Mean	185	196	210	200
Median	180	229	188	207
COD	Conc.	Loads	Conc.	Loads
Value	mg/l	kg COD/d	mg/l	kg COD/d
Nos of observ.	23	19	14	14
Mean	495.5	451.2	681.5	672.1
Median	403.0	361.5	526	616

Pollution Ratios in Raw Wastewater

A way to assess the used analysis reliability is to calculate the pollution ratios in the wastewater. The ratios on relevant pollution variables on untreated wastewater are also compared with what is assumed to be a "typical municipal wastewater". In the following ratios are given for two operation periods, September 2011 through April 2012 (first period), and May 2012 (second period):

Table 5: Summary of total nitrogen loads into Ja-Ela WWTP January 2012 through April 2012, and May 2012

Period	January-April 2012			
Nos of observ.	23	nos	19	nos
Mean	57.4	mg/l	58	kg N/d
Median	51.3	mg/l	50.3	kg N/d
Period		May 2012		
Nos of observ.	15	nos	14	nos
Mean	68.1	mg/l	64.8	kg N/d
Median	64	mg/l	60	kg N/d

Table 6: Summary of total phosphorus loads into Ja-Ela WWTP January 2012 through April 2012, and May 2012

Period	January-April 2012			
Conc.			Loads	
Nos of observ.	10	nos	7	nos
Mean	8.2	mg/l	9.0	kg P/d
Median	8.0	mg/l	8.2	kg P/d
Period		May 2012		
Nos of observ.	Conc.		Loads	
Mean	15	nos	14	nos
Median	7.2	mg/l	7.2	kg P/d

First period: COD: BOD, mean value of 23 observations = 2.5:1; this value coincides with a "typical municipal wastewater"; Second period: COD: BOD, mean value of 4 observations = 3.3:1; this value higher than the "typical value for municipal wastewater". The number of observations in this case is low, thus the earlier observation regarding the previous operation period is deemed far more reliable.

First period COD: SS, mean value of 77 observations = 4.1:1; this value is higher than a "typical municipal wastewater"; Second period: mean value of 10 observations = 1.2:1; this value is lower than a "typical value for municipal wastewater".

First period COD: total N, mean value of 79 observations = 8.7:1; this value is slightly lower than a "typical municipal wastewater"; Second period COD: total N mean value of 10 observations = 11.3:1; this value is in accordance with "a typical municipal wastewater".

First period: COD: total P, mean value of 32 observations = 99.6:1; this value is higher than a "typical municipal wastewater". Second period mean value of 10 observations = 88.3:1; this value higher than the "typical value for municipal wastewater".

Discharge Levels from the Plant

The organic pollution discharge from the plant is measured mainly as COD, although also the BOD_5 also is presented with few observations.

In Table 7 is shown the discharge concentrations for BOD_5 and COD during the period January 2012 through mid-April 2012 and for May 2012. Comments to Table 7: The discharge levels of BOD_5 are all substantially lower than the ruling consent value = 25 mg/l. As the plant is running with very low loading and a virtually complete nitrification, see the figures below, the discharge levels of mainly around 5 mg/l are expected. It may also be observed that the discharge figures in May 2012 have been significantly lower than during the first trimester of 2012. The discharge levels of COD are all substantially lower than the ruling consent value = 120 mg/l. As the plant is running with very low loading and with possibly a very high SRT (Solids Retention Time) the results are as expected, with normal discharge levels <50 mg COD/l. It may also be observed that the May results are significantly lower than the values during the first trimester 2012. The recorded removal levels of COD suggest that the incoming organic pollution is mostly comparatively easily degradable.

Table 7: Summary of BOD_5 and COD discharge levels from Ja-Ela WWTP January 2012 through April 2012, and for May 2012

period	January-April 2012	May 2012	
Nos of observ.	14	8	nos
Max value	11.0	4	mg/l
Mean	5.5	2.1	mg/l
Median	4.05	2.0	mg/l
Min value	2.00	1.0	mg/l
Standard deviation	2.98	1,0	mg/l
Standard error	0.80	0.4	mg/l
COD, Obs period	January-April 2012	May 2012	
Nos of observ.	26	15	nos
Max value	73.1	43	mg/l
Mean	48.1	35.9	mg/l
Median	45.6	36.0	mg/l
Min value	37	28.4	mg/l
Standard deviation	9.13	4.0	mg/l
Standard error	1.79	1.0	mg/l

Discharge of Nutrients from the Plant

In a similar manner the nutrients, nitrogen and phosphorus are recorded during the period. Also for these variables only the 2012 year figures are presented. In the case of nitrogen, total N, ammonia N and nitrate N are analyzed and recorded. The nitrogen discharge is presented in Table 8 (total nitrogen, ammonia nitrogen and nitrate nitrogen). Comments to Table 8: The fact that the plant is operated at very low load levels is reflected in the nitrogen removal performance. Generally the removal is found to be very good with discharge levels of total nitrogen at around 10 mg N/l compared with the ruling consent level of <30 mg N/l. The nitrification level is very high with a median value of 1.28 mg NH_4-N/l in effluent, compared with the ruling consent level of <10 mg NH_4-N/l. The discharge of nitrate is found to be around 5 mg NO_3-N/l. According to the results it is estimated that the discharge level of organic nitrogen is around 3 mg N/l. This in turn may indicate that the industrial influence is important. A well nitrified municipal wastewater would have an organic nitrogen discharge content of around 1 mg N/l at similar inlet nitrogen concentrations.

Table 8: Summary of nitrogen compounds in effluent from Ja-Ela WWTP January 2012 through April 2012, and May 2012

Period	January-April 2012			
Nitrogen compounds	Total N	NH4-N	NO3-N	
Nos of observ.	23	10	9	nos
Max value	27.8	4.5	8.0	mg/l
Mean	10.4	2.0	5.0	mg/l
Median	9.1	1.3	4.8	mg/l
Min value	5.5	0.43	2.9	mg/l
Standard deviation	4.78	1.55	1.56	mg/l
Standard error	1.00	0.49	0.52	mg/l
Period			May 2012	
Nitrogen compounds	Total N	NH_4-N	NO_3-N	
Nos of observ.	15	13	13	nos
Max value	11.8	3.5	3.4	mg/l
Mean	7.2	1.7	2.1	mg/l
Median	6.9	1.6	1.9	mg/l
Min value	5.1	0.1	1.1	mg/l

| Standard deviation | 1.5 | 0.9 | 0.8 | mg/l |
| Standard error | 0.4 | 0.3 | 0.2 | mg/l |

Comments to Table 9: The number of observations is in this case rather limited. One observation shows a high effluent level, while the remaining observations are under the ruling consent value of 5.0 mg P/l. However, in the light of the overall performance and the adopted process chain it would be possible to run the plant with lower effluent P values. The performance during May 2012, after the introduction of an intermittent aeration mode has resulted in a quite substantial improvement of the phosphorus removal.

DISCUSSION

The Ja-Ela WWTP is currently running with very low inlet loads in comparison with the design data. This is true for both flows and the major relevant pollutants. The overall treatment results are far better than the ruling consent levels. During the ruling circumstances anything else would be more than astonishing. However, the low load situation in relation to the reactor volumes created a number of set-backs that had to be mitigated as the initial operation caused some efficiency problems. The major ones are summarized as follows:

- The operation, until mid-April 2012, has been based on a continuous aeration in the bio-reactor aimed for nitrification. This has resulted in unnecessarily high free oxygen levels, and as a consequence, an excessive energy use for aeration. When the aeration has been operated 24 hours a day the consumed power for aeration then is around 1440 kWh/d. The matter was also reflected by the free oxygen level in the nitrification redactor being 5 and 7 mg O_2/l at the outlet part of the aeration reactor;
- It should be observed that the aerobic reactor has a wet volume of 2383 m³. Thus the hydraulic retention time as an average is more than 2 days at the prevailing conditions;
- The currently operated solids retention time (SRT) is reported to be very high by the Contractor. As only small amounts of excess activated sludge have been withdrawn from the reactors it is very likely that this observation is correct;

- The return activated sludge rate is also very high, the ratio during day hours is roughly 10:1, while a far more realistic ratio would be in the order 2:1 to 0.75:1;
- The current operation mode with too high free oxygen levels (also supported by the high recirculation rate of return activated sludge), very high SRT and complete nitrification may well promote the rather high discharge levels of phosphorus. Although this matter is not critical it may be possible to lower this discharge level by process modifications.

Table 9: Summary of total P, Conc. in effluent from Ja-Ela WWTP January 2012 through April 2012, and May 2012

Period	January-April 2012	May 2012	
Total P			
Nos of observ.	11	15	nos
Max value	8.0	3.5	mg/l
Mean	4.6	1.2	mg/l
Median	4.1	1.0	mg/l
Min value	3.4	0.6	mg/l
Standard deviation	1.42	0.7	mg/l
Standard error	0.43	0.2	mg/l

Based on these considerations some alterations were made in the operation strategy, starting at the beginning of May 2012. The most apparent observations are the following:

- The new operation mode based on an intermittent operation of the aeration has provided an improved treatment result. The following operation scheme has been implemented from the last day of April 2012: This includes a typical "operation cycle" with a total length of 1.8 hours, thus giving a total number of cycles = 12/d. The total aerated time is 3.5 h/d and the stop time 20.5 h/d.
- Generally speaking the change of operation mode has been successful with respect to the supplied oxygen to the reactor. Now, based on this scheme it is possible to calculate the typical, specific load values for F/M and nitrogen in the aeration basin. Observe that in this case only the aerated time is used as a basis

for an "aerobic F/M, based on COD" and the potential N load related to nitrification. The mean value for the aerobic F/M has been 0.61 kg COD/kg SS/d and the specific aerobic nitrogen load has been 0.053 kg N/kg SS/d during the altered operation in May 2012.

- The sludge volume (SV) is analyzed on a regular basis and thus it is possible to calculate the SVI (Sludge Volume Index). In Table 10, a statistical presentation of SS levels in the bioreactors, the measured SV and the calculated SVI are presented.

The sludge quality data are so far deemed to be very good, with SVI levels normally <90 ml/g. The currently operated solids retention time (SRT) is reported to be high by the Contractor. As only small amounts of excess activated sludge have been withdrawn from the reactors it is very likely that this observation is correct. In order to assess the SRT in a quantitative manner it is important to measure the SS content in the bioreactor system, in the waste activated sludge system and in the discharge wastewater. It is also imperative to measure the waste activated sludge flow and record the amounts of discharge wastewater.

Some "shortcomings" in the applied design in relation to an intermittent operation are worth-while to observe. The wasting of activated sludge is done from the return activated sludge pipe. Thus it has been difficult to control the SRT in the best possible way during the intermittent operation. The automation system allows only for a maximum stop of the blowers by 99 minutes, thus a further optimization has been limited. The cooling of the blowers has also become a limitation. This is mostly related to the prevailing air temperature of 25°C - 35°C.

Table 10: Summary of sludge quality data in bioreactors at Ja-Ela WWTP January 2012 through April 2012, and May 2012

Period			January-April 2012	
Variable		SS	SV	SVI
		mg/l	ml/l	ml/g
Nos of observ.		33	34	33
Max value		4155	350	106
Mean		2081	154.4	71.3
Median		1944	125	65.6

Min value	940	52	35.9
Standard deviation	590.34	80.64	23.50
Standard error	102.77	13.83	4.09
Period		May 2012	
Variable	SS	SV	SVI
	mg/l	ml/l	ml/g
Nos of observ.	14	15	15
Max value	4400	350	93
Mean	3787	316	86.2
Median	3810	320	86.2
Min value	3198	280.0	80.0
Standard deviation	373	22.0	3.7
Standard error	99.6	5.7	1.0

CONCLUSIONS

The issue of operation of an under-loaded activated sludge treatment facility has been addressed and the basic problem has been identified as an excessive use of energy for the aeration. The way to mitigate this problem showed to be a very simple one—to operate the aeration basin at an intermittent mode. In this specific case it has been possible to implement this model; however, some important limitations have been identified at the same time:

- It is found imperative that the on-line measurement of operation variables are viable and maintained on a regular basis;
- The automation mode must allow for a flexible intermittent operation. This is a consideration that should be reflected already in the design work;
- An additional removal strategy for the waste activated sludge may be needed: To withdraw the sludge from the intermittent reactor;
- The current mode of intermittent operation has been successful also thanks to an extended hydraulic retention time in the aerobic reactor. This matter must be addressed closely when applying the strategy; at shorter hydraulic retention times there is a risk that amounts of non-nitrified ammonia will pass through the system;

- The outcome of the operation modification has been by large very satisfying, with a sustained biological performance with respect to organic removal (expressed as BOD and COD), an improved removal of total nitrogen thanks to an efficient denitrification and finally but not least striking: An efficient biological phosphorus removal has been established in the vicinity of 80% to 90%.

ACKNOWLEDGEMENTS

The client's personnel, represented by H. A. Ariyiadasa, Head of Wastewater Treatment Plant NWS&DB, and G. D. N. Neville, Chief Project Engineer, NWS&DB; R Kulanatha, Project Director NWS&DB have been helpful at the plant—providing viewpoints and information on the day-by-day operation. Mr. Staffan Indebetou at Purac has given valuable information on the plant design details. For the linguistic check-out Mr. Guy Taylor has been of great help.

REFERENCES

1. M. J. Tetreault, B. Rusten, A. H. Benedict and J. F. Kreissl, "Assessment of Phased Isolation Ditch Technologies," The 59th Annual Conference of the Water Pollution Control Federation, 7 October 1986.
2. R. I. Irvine, "Technology Assessment of Sequencing Batch Reactors," US Environmental Protection Agency, Cincinnati, 1983.
3. R. L. Irvine, et al., "Analysis of Full-Scale SBR Operation at Grundy Center, Iowa," Journal of Water Pollution Control Federation, Vol. 59, No. 3, 1987, pp. 132-138.
4. E. C. Hoepker and E. D. Schroeder, "The Effect of Loading Rate on Batch-Activated Sludge Effluent Quality," Journal of Water Pollution Control Federation, Vol. 51, No. 2, 1979, pp. 264-270.
5. S. Marklund and S. Morling, "Biological Phosphorus Removal at Temperatures from 3 to 10°C—A Full Scale Study of a Sequencing Batch Reactor Unit," Canadian Journal of Civil Engineering, Vol. 21, No. 1, 1994, pp. 81-88. doi:10.1139/l94-008

6. L. Rieger, I. Takás and H. Siegrist, "Improving Nutrient Removal while Reducing Energy Use at Three Swiss WWTPs Using Advanced Control," Water Environment Research, Vol. 84, No. 2, 2012 pp. 170-188. doi:10.2175/106143011X13233670703684

Mineral Industry in Egypt-Part I: Metallic Mineral Commodities

Abdel-Zaher M. Abouzeid[1] and Abdel-Aziz M. Khalid[2]

[1]Department of Mining, Petroleum, and Metallurgy, Faculty of Engineering, Cairo University, Cairo, Egypt
[2]Geological Survey and Mineral Resources Authority, Cairo, Egypt

ABSTRACT

This mineral potential in Egypt is quite high. Almost all sorts of industrial minerals such as metallic and non-metallic commodities exist in commercial amounts. However, Egypt imports many of the mineral commodities needed for the local mineral industries. The main reason for this is that the investors, either the governmental or the private sectors, refrain from investing into the mineral industry for prospecting, evaluation, and developing the mining and mineral processing technologies. This is because the return on investment in the mining industry is generally low and the payback period is relatively long compared with easy-to-get money projects. Another reason is the disarray of the mining laws and regulations and lack of administrative capability to deal with domestic and international investors and solve the related problems. Also, lack of skilled personnel in the field of mining and mineral processing is an additional factor for the set back of the mining industry in Egypt. This is why the mining technology in Egypt is not very far from being primitive and extremely simple, with the exception of the underground mining of coal, North of Sinai, and Abu-Tartur phosphate mining, where fully automated long wall operations are designed. Also, the recent gold and tin-tantalum-niobium projects are being designed on modern surface mining and mineral processing technologies. The present review presents an overview of the most important metallic mineral commodities in Egypt, their geological background, reserves and production rates. A brief mention of the existing technologies for their exploitation is also highlighted.

INTRODUCTION

Egyptian Civilization is one of the most ancient civilizations in the world, which practiced mining and processing of metallic and non-metallic ores. The ancient Egyptians quarried the dimensional stones in a very orderly manner to obtain geometrically shaped blocks with exact dimensions to build tombs, temples and pyramids. They also cut-from extremely hard rocks such as granite, gab-bros, and granodiorites-obelisks and blocks for hewing statues and for recording their history on them. They also traced the natural minerals, collected them, and treated them to compose the ever-beautiful painting colors, which stayed bright and persisted weather changes for thousands of years. The Ancient Egyptians had an excel-lent sense and knowledge about geology, survey, rock mechanics and metallurgical processing. They worked their way out in open pits, open cast, and underground mining. Almost all gold and copper locations known at present were originally discovered and worked out by the Ancient Egyptians. The technology limitations in mining, and processing, at that time, limited the mining depth, and the overall efficiency of upgrading the ores. The first known underground map (1300 BC), for El-Fawakhir gold mine, is preserved in Turin museum in Italy.

There are evidences that the Ancient Egyptians mined and extracted gold, silver, copper, and zinc. They used these metals in their pure state and/or as alloys to suit certain purposes. They designed and produced several hard alloys such as bronze (90% Copper and 10 % zinc). They also traced all sorts of gem stones in Sinai, Eastern Desert, and Western Desert. They quarried lime-stone, granite, marble, breccias, diorites, and granodiorite stones.

Mining in Egypt today, follows almost the same meth- odology as the Ancient Egyptians used to use thousands of years ago. The main differences are in the introduction of the modern technologies which are available today and were not available then. The underground mines to-day are much deeper, drainage of the underground water is readily drained by means of pumps which were not available at that time, the underground atmosphere is conditioned by the up to date conditioning techniques (ventilation and refrigeration), the underground openings are electrically lightened, and the raw materials are mechanically transported [1]. However, the scale of mining in Egypt at present is still small. The largest mining operation, which is the iron ore mining, does not exceed 3 million ton/year [2].

Geological Background

The Egyptian territory is covered by crystalline basement rocks belonging to Precambrian age and Phanerozoic sediments which range in age from Cambrian to Recent. The basement rocks form about 10% of the land surface and are exposed in South Sinai, Eastern Desert, and South West corner of Egypt [3].

The basement rocks of Southwest corner of Egypt crop out from Egyptian-Libyan borders to Gabal Kamel as continuous low land and ridges. From Gabal Kamel to Aswan, the basement occurs as uplifting inliers. Richter [4] and Richter and Schandelmeier [5] classified Precambrian rocks into three Formations starting from High grade granulites (Granoblastic Formation), overlain by the remobilized Anatexite Formation, and finally the youngest clearly bedded Metasedimentary Formation. All these formations were intruded by granodiorite and porphyritic granite. This classification was used by EGPC and CONOCO [6] in issuing a map of a scale 1:500 000 for the area. Naim et al. [7] used a simpler classification system, where they classified these rocks as old metamorphic rocks of probably Archean age to early Proterozoic. Klerkx [8] and Sultan et al. [9] included amphibolite, ortho- and para- gneisses of granulite and amphibolite facies, which are intruded by calc-alkaline granitoids and gabbros. The late magmatic rocks are clearly related to Pan African. The main economic min-erals in this region are Banded Iron Formation probably of Lake Superior Type [10-12].

The basement rocks of Eastern Desert and Sinai form part of Arabian-Nubian shield. More than one scenario were proposed for the evolution of this shield, the more acceptable one is that which assumed that the shield is built up of arc(s)-inter arc(s) rock association [13,14]. The arc associations' complexes encompass the volcano- sedimentary group. The arc-inter arc associations are well illustrated by ophiolitic slabs (serpentinized ul-tramafic rocks, tholeiitic Meta gabbros, mafic meta-volcanics), thrusted over the arc complex terrain [15]. The evolution and cratonization of the arc group took place between 900-550 Ma [16]. The arc-inter arc group was intruded by syn-to late- tectonic calc-alkaline dio-rite-granodiorite rocks through tectono-magmatic cycle ended by cratonization, through thrusting, low angle shearing and folding [17]. This stage was culminated by granodiorite intrusion as in Meatiq and Hafafit areas at 612 Ma [18]. The Neoproterozoic crust was subjected to regional NW-SE folding and intruded by granite, G1 [19] and Dokhan volcanics [20]. Hammamat sediments which are derived from volcanic rocks were deposited in intra mountain basins [21]. The tectonic granite (younger granite G2 and G3) [19] was intruded in the final stage of Pan African tectonomagmatism [20]. During rifting stage several ring complexes were intruded as intraplate magmatism which could be emplaced during Paleozoic or younger [22]. During Phanerozoic, three-within-plate volcanic activities namely Katherine Volcanic, Wadi Natash Volcanic, and Tertiary basalts and dolerites are recorded.

The Phanerozoic sediments overlay uncomfortably the basement rocks and cover 90% of the whole Egyptian territory starting from Paleozoic to Quaternary.

Older Paleozoic rocks crop out near the basement contacts in the Western Desert and in Sinai, but they sink below younger sediments further North and West. In Sinai the section is mostly sandstone usually ferruginous and manganiferous in Um Bogma [23].

Mesozoic sediments are very unequally distributed. Marine Triassic is found in Aref El Naga, whereas continental covers many areas in Egypt. The Jurassic age is well developed in Gabal Maghara and South West of Sinai, Khashm El Galala. Cretaceous sediments are widely distributed and form about 40 % of the Egyptian surface. The deposition of Cretaceous sediments is not only governed by regression and transgression of fall and rise of sea level, but also by renewal uplift of source areas and variations in continuous input linked to tectonism along the continental margin [24] and within the craton [25]. All phosphate deposits and white sands are the economic minerals in Mesozoic.

The Cenozoic in Egypt witnessed three major events: thin distribution in time and space, their mode of deposition, and N-E and S-W changes of facies [23]. The Pa-leocene and Eocene rocks crop out in the Nile Valley between Luxor and Cairo, Fayioum, Bahariya, Sinai, and North Eastern Desert. Oligocene started with the uplift-ing of Egyptian Craton with the rise of South Western Desert of Egypt. It is well illustrated in Gabal Qatrani, Gabal Ahmar, and Safaga-Quseir area and around Mersa Alam on the Red Sea, Baharia, and West of Nile Valley. Miocene occupies one eighths of the total land surface [26]. Sedimentation was greatly influenced by the tec-tonic events which led to the formation of the Red Sea [23]. 1t crops out at Red sea, South West of Sinai and North Western Desert. Pliocene is represented in Red Sea hills, Wadi Qena, Mediterranean Sea, Fayoum, Nile Val-ley and Nile Delta, and Cairo-Suez district. Quaternary deposits are widely distributed as wadi deposits, sand dunes, and Sabkhas.

Economic Metallic Ores in Egypt

Several metallic ores were recorded in Egypt [27]. In the present time, only iron and ilmenite are under mining while manganese and chromite are mined in small scale. The rest of metallic ores mainly, gold, Pb-Zn, Cu, Nb-Ta deposits are still under exploration and re-estimation of ore reserves.

Many attempts were done to classify these ores either on the bases of time of deposition [28-30] or in the frame of metalogenetic aspects [31, 32]. The first linking be-tween plate tectonic modeling for Arabian-Nubian shield and mineralization was given by Garson and Shalaby [33]. The latest classification was proposed by Botros and Noor [34] where they classified the Egyptian ore deposits on the bases of tectonic-magmatic stages as follows;

Island Arc Stage

- Deposits formed in ophiolitic assemblage including Cu-Ni-Co sulphides e.g. Abu Swayeil copper and Podi-form chromite deposits.
- Deposits formed in primitive island arc including Banded Iron Formations, BIF, and its gold related deposits.
- Deposits formed in mature island arc including volcanic hosted base metal massive sulphides e.g. gold related deposits such as Um Samuki.

Accretional Stage (Orogenic Stage)

- Auriferous vein type.
- Base metal vein type.
- Titanoferrous iron ore, e.g., Abu Ghalqa ore deposit.

Late Orogenic-Extensional Stage

- Cu-Ni sulphides Gabbro, Akarm
- Titanoferrous iron ore, Kurabkanci
- Association with granitic rocks:

  -Beryllium, e.g., Um Kabu

  -Tin –deposit e.g. Abu Dabbab

  -Tungsten, e.g., Igla

  -Fluorite, e.g., Homr Akarm

  -Auriferous vein deposit, e.g., El Sid

This series of articles provides a statistical summary about the most important mineral commodities in Egypt. It also briefs the geological aspects, the mining, and min-eral processing techniques used in the today's mining activities and the scale of the mining operations existing in Egypt. Each commodity is preceded with the related geological conditions and events. The mineral commodi-ties can be classified as metallic and non-metallic deposits [1, 2, 11, 35-41]. The most important of these deposits are:

- Metallic ores such as: iron ores, gold ores, industrial metal oxides (Sn, Ta, Nb, W, and Mo), titanium and titaniferous-iron ores, manganese ores, sulphide mineralization (Pb, Zn, Cu, and Co), and chromite.
- Non-metallic ores such as: phosphate, limestone, dolomite, ornamental stones, quartz rock, white sands, talc, feldspars, kaolin, fire clays, bentonite, gypsum, fluorspar, sands and gravels, magnesite, evaporates (salts), and coal.

The present work is an attempt to shed some lights on the metallic ores in Egypt as a whole and to discuss the technological problems facing their exploitation, i.e., no specific mineralization classification will be strictly followed. The metallic ores, which will be discussed here in, are put according to the priority of their economic impact on Egypt.

IRON ORES IN EGYPT

Iron ores in Egypt occur in two forms:
- Banded Iron Formation (BIF), and
- Ironstone.

The iron deposits in Egypt are shown in Figure 1 [27]. This figure shows the distribution of iron ores and iron oxide traces all over Egypt. Most of the locations are inter-related in origin to each other. The trend of the iron oxides in Western Desert points out to a common source of the iron deposits in this area.

Banded Iron Formation, BIF

This type of iron mineralization was recorded in Eastern Desert in 13 localities between Safaga in the North and Mersa Alam in the South, and in South Western Desert in the area between Egyptian-Libyan border in the West to Gabal Kamel in the East and extends outside Egypt to the Libyan and Sudanese territories around Gabal Arkenu and around Gabal Kissu, respectively [42].

Eastern Desert BIF

The most famous occurrences of BIF in Eastern Desert are Wadi Kareim, Um Nar, Abu Marawat, El Dabbah, Um Ghamis, Gabal El Hadid, Um Shadad, and Abu Di-wan. The country rock of BIF is the island arc volcanic (basalt, andesite, and dacite), and volcanoclastic rocks as in Abu Marawat [43]. The volcanoclastic rocks are ac-cumulated with BIF in intra-arc basins and intercalated with BIF in the central part of Eastern Desert. The di-mension of BIF band ranges in thickness from few cen-timeters up to 5 meters with an average thickness of 1.5 m in most cases.

Um Nar BIF is a good example, where this area has 13.7 million tons with iron content up to 45 % Fe [32]. According to Dardir and El Chimi [44], this area is mainly built of four litholotectonic units: 1-porphyritic quartz granitoids, structurally overlaying a series of tectonically mixed schists comprising acid and intermediate tuffs, biotite, quartz, schists and phyllonite, and Serpentine which forms Gabal El Mayit ultramafic rocks. Um Nar sequences overlie structurally the serpentine rocks of Gabal El Mayit. This group is folded into overturned syncline, Figure 2 [45]. Figure 2 shows the complexity of the iron formation at Um Nar and similar iron ore lo-cations in the Eastern Desert. There are extensive folding and faulting systems in the area. This structure reflects the difficulties which may face the mining operations. BIF in all outcrops exhibits oxide facies which is com-posed of alternative iron rich bands with silica rich bands. Carbonates and sulphides facies also do exist [46, 47]. El Dougdoug *et al.* [47], on the bases of the mineralogy of Gabal El Hadid, stated that BIF exhibits the following formations: 1-hematite-magnetite-jasper as oxide facies, 2-siderite-magnetite-chert as carbonate facies, and 3- pyrite-magnetite as sulphides facies. Regarding the origin of BIF, Botros [43] proposed a model for Abu Marawat BIF, where he attributed the formation of BIF to the interaction between volcanically derived fluids and sea water. These fluids were capable to leach iron, silica, and other associated elements including gold from basalt and andesite in early stage of island arc volcanicity (immature island arc).

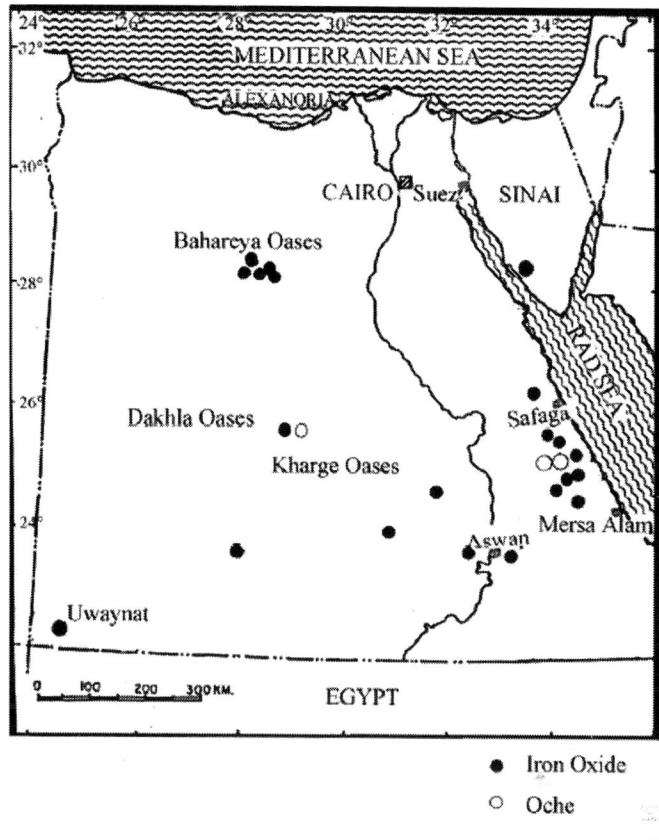

Figure 1: Locations of iron mineral deposits in Egypt [27].

Western Desert BIF

BIF in Western Desert was discovered in two main areas namely Gabal Nazar and Gabal Kamel. At Gabal Nazar area, which lies just to the East of the Libyan borders, the banded iron formation occurs as thin bands (5-10 m) within amphibolite and quartzofelsphathic gneisses [12, 38] And takes E-W and NE-SW trend. From economic point of view, it is less important due to the intrusion of huge granitic bodies which cut the extension of the ore. Khalid and Diaf [12, 38] recorded some gold anomalies in this formation. At Gabal Kamel area, BIF was recorded in the area between latitude 22°00' to 22°20' north and longitude 25°30' to 26°40› east, which are covered by Archean to early Proterozoic rocks [9]. These metamorphic rocks include garnet-granulite, mafic granulite, para-and orthogenesis showing intense deformation manifested by folding and faulting with at least four de-formation phases [48]. The whole area had undergone regional metamorphism from garnet granulite to amphibolite facies [4]. Anatexis features are well developed in the area.

The BIF occurs as strongly folded and faulted bands within these Archean to Early Proterozoic high grade metamorphic sequence. The formation attains 300 m thickness and extends several kilometers in length, Figure 3 [7, 11, and 39]. This figure shows the patchy formation of the iron oxide blocks. These blocks are scattered vertically and laterally. Mining of iron oxides in these areas will be extremely difficult, and removal of the huge inter-bedded quartzite and quartz will be highly costly. It will require highly advanced selective mining techniques.

BIF in this area is classified according to the mode of occurrence into three types: 1-well banded type, 2-brecciated type, and 3-ferruginous chert [39]. The well banded type is the predominant where iron rich bands (magnetite, hematite, and goethite) alternate with micro-crystalline silica rich bands (quartz, chert, and jasper). Mineralogical studies revealed that opaque minerals are magnetite, hematite, and goethite which are the main minerals with some sulphides (pyrite, chalcopyrite, arsenopyrite, and covellite). Graphite was recorded in some samples which could be attributed to metamorphism of carbonate facies [39]. The chemical analyses show that iron oxides range between 16% and 55.5% [39]. On the bases of geographic situation and major structural elements and mineralogical characteristics, Khattab *et al.* [11] classified the BIF in this region into Western, Central, and Eastern zones. They came to a conclusion that the Central and Western parts are, economically, the most promising areas. All the above mentioned authors are inclined to consider this type of BIF as Lake Superior Type of oxide facies and suggested deposition in epicontinental marine basin with free access to the ocean [49].

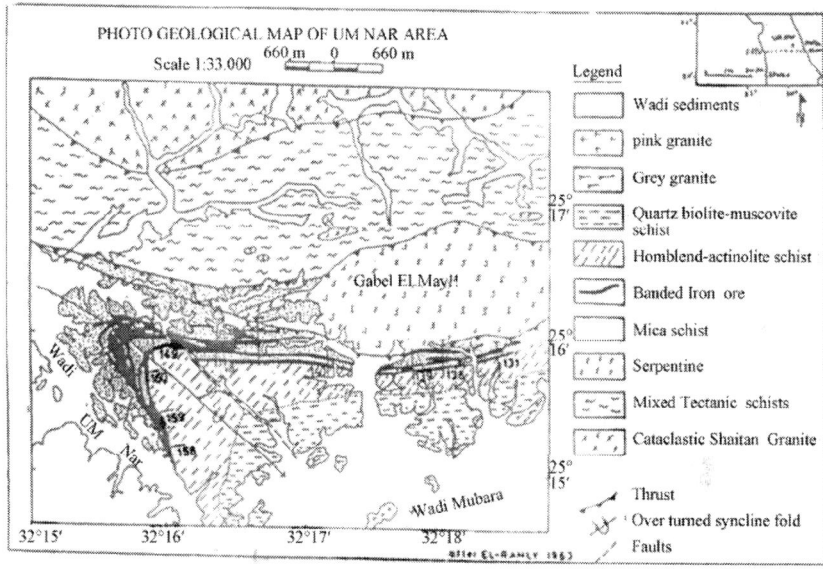

Figure 2: Geological map of Um Narbanded iron formation, Eastern Desert, Egypt [45].

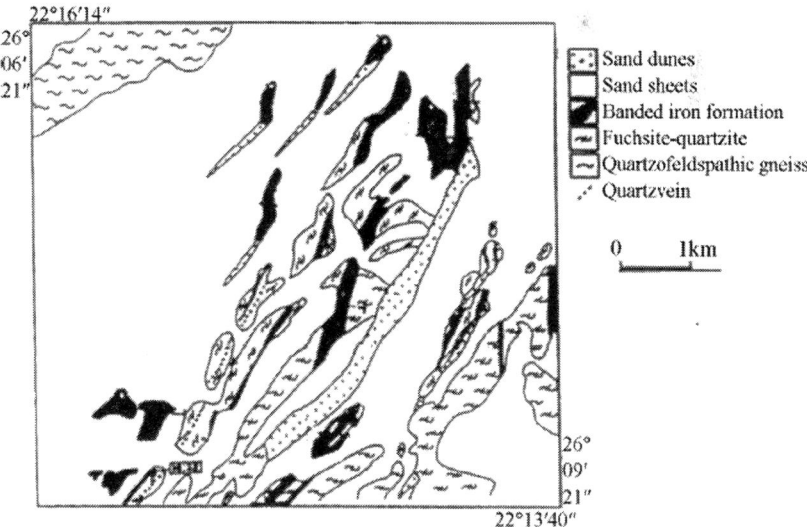

Figure 3: Geological map of Area K 7 in Central Zone of Gabal Kamel BIF [11].

Ironstone Deposits

Iron stone is the iron ore which is formed mainly within Phanerozoic sediments and is well represented in Egypt in Bahariya and Aswan iron ores.

Bahariya Iron Ore: Several iron ore deposits are lo-cated in Bahariya area, e.g., El Harra, El Heiz, Ghorabi, El Gedida, and Nasser. The iron ore of El Harra belongs to El Harra member of El Haffuf Formation; whereas El Gedida iron ore belongs to Naqb Formation [50-52]. The area is covered by Bahariya Formation (unfossiliferous varicolored sandstone of Cenomanian age) followed by El Heiz Formation (brownish limestone and sandy clay beds), and El Haffuf Formation of sandstone, sandy clay, and ferruginous beds, which are partly taken by the iron ore deposit, Khuman Formation (chalky limestone), and Naqb Formation of thick limestone beds with few marl and clay associations. The iron content in the ironstone deposits ranges from 30% to 58% Fe, and the manganese content ranges from 0.7% to 7.66% Mn [52].

The stratigraphic position of Naqb Formation is partly taken by iron ore deposits at El Gedida, El Harra, and Ghorabi; where El Gedida iron ore member belongs to iron deposits of Lower Middle Eocene (Naqb Formation) and the upper Eocene (Abu Maharik Formation. The ore is localized in the crest of anticline [53].

Origin of Ironstone Ores at Bahariya Oases: The origin of the existing Ironstone ores was discussed by several authors [52-58]. El Shazly [59] assumed that the Ghorabi iron ore was derived from the chemical weathering of older rocks. El Bassyouny [52], on the bases of detailed field work, stated that the iron content of El Harra ironstone deposit increases in iron content towards the fault and decreases gradually northward, away from the fault, where ferruginous limestone crops out, and he believes that the enrichment of iron took place by intense metasomatism replacement which is believed to had taken place in post middle Miocene–Eocene time, probably related to nearby volcanism. On the other hand, El Aref and Lotfy [55] proposed karst genetic for El Bahariya iron ore, where they suggested that the iron depos-it's were formed through lateritization processes during the senile stage of post Eocene karst event. Karst depressions and excavated unconformity acted as traps where iron oxides are accumulated. Iron deposits together with soil products also form surfacial crust (duricrust), cap-ping and cementing highly subdued and altered carbonate rocks. The evolution of megascopic and microscopic ore fabrics, the oxidation of iron bearing minerals, and their relation to the gangue and weathering products reflect the changes in the moisture regimes and the physicochemical conditions involved during the pedogenesis [55]. Hussein [32] proposed a very important idea, where he considered that most of the folds were generated by faulting affiliated with the Pelsuium mega-shear along which the Bahariya Oases are located [58]. The present authors are inclined to believe in this idea where along this zone iron was recorded by Issawi [59] in Black Hills of high iron content (up to 38% Fe). More iron discoveries are expected along this zone especially to the Southward direction, where the main iron source is Basement rocks (Banded Iron Formation).

Figure 4, shows a map for Bahariya Oases with its iron ore localities as related to each other geographically only the exploitable iron ore in Bahariya Oases at El Gedida area, which has little or no overburden. Originally, when mining started in this area in 1972, the minable reserves were estimated accurately by 135 Mt. Today, the left minable reserves are estimated by only 63 Mt, which are just enough for about 15-20 years at the present mining rate of 3 to 3.5 Mt/y The other areas:

Ghorabi, Nasser, El Heiz, and El Harra are of low grade ores and of high manganese content. In addition, these areas have relatively thick overburden.

Aswan Iron Ore: Iron ore in this area was known since Pharaonic time. In recent years it was the main supply of iron ores for the Egyptian iron and steel industry till 1972 when it was replaced by Bahariya iron ore. According to Hussein [32], the ore is a bedded oolitic type of Senonian age in the form of two bands interbedded with ferruginous sandstone and clay capping Precambrian rocks. The thickness of the bands varies from 0.2 to 3.5 m. The main iron minerals are hematite with minor goethite where quartz, gypsum, halite and clay are gangue miner-als. The reserve was estimated between 121-135 million tones with average content of 46.8% Fe [60]. The ore had been formed under sedimentary lacustrine conditions during Senonian sedimentation. Aswan iron ores were used to feed the steel plant at Helwan from its establishment in 1956 until 1972 when the Bahariya iron ore, from El Gedida area, started to replace Aswan ore in the iron and steel plant at Helwan.

The potentiality of discovering more iron ore is high especially in the area between Bahariya Oases in the North and Uwaynat area in the South on the bases of geological and structural observations. Reserves and production of iron ores in Egypt are shown in Table 1. In this table, it is clear that Egypt is running short of the available indigenous exploitable iron ores, which is mainly in El Gedida area, Bahariya Oases, Western Desert. The Aswan iron ore is high in phosphorus content, in addition to its peculiar formation. The Eastern Desert BIF iron ores are of small quantities (about 50 million tons in total in all localities) spread in a vast area of about 200 km2. Uwaynat BIF iron ore has been recently discovered and not thorough investigations (exploration, reserve estimation, characterization and /or beneficiation) have been carried out. In addition, the area is more than 600 km far from the inhabited Nile valley area, with little or no infra-structure. A preliminary project for mining, processing, and pelletizing of Uwaynat iron ore deposit is being proposed after the discovery of Uwaynat BIF

Deposits [17, 38, and 48] Pilot scale mineral processing tests showed that Uwaynat ore may be upgraded to 66 % Fe at a reasonable recovery [39].

The mining method in all producing locations at pre-sent is open pit mining [61]. The main iron ore processing plant is at Bahariya Oases (El-Gedida Area) [61]. The plant consists of a jaw crusher followed by a cone crusher to reduce the run-of-mine ore to the maximum size required by the sinter plant at Helwan. Of course, this is waste of energy and cost. The raw ore is trans-ported, with all its gangue content, from the mine to the steel plant for a distance of over 300 km. This represents higher transportation costs, loss of energy in extraction, lower unit productivity, waste of labor efforts, and so on. It could have been more beneficial if the ore is concentrated in the mine site, i.e., raising the iron content of the ore from 52 % Fe to 65 % Fe. This will overcome all the above drawbacks of using the mined ore as it is. There are modern technologies for upgrading such type of ores. These technologies include flotation, flocculation/flotation, high intensity magnetic separation, and magnetic roasting followed by low intensity magnetic separation.

Mineral Industry in Egypt-Part I: Metallic Mineral Commodities

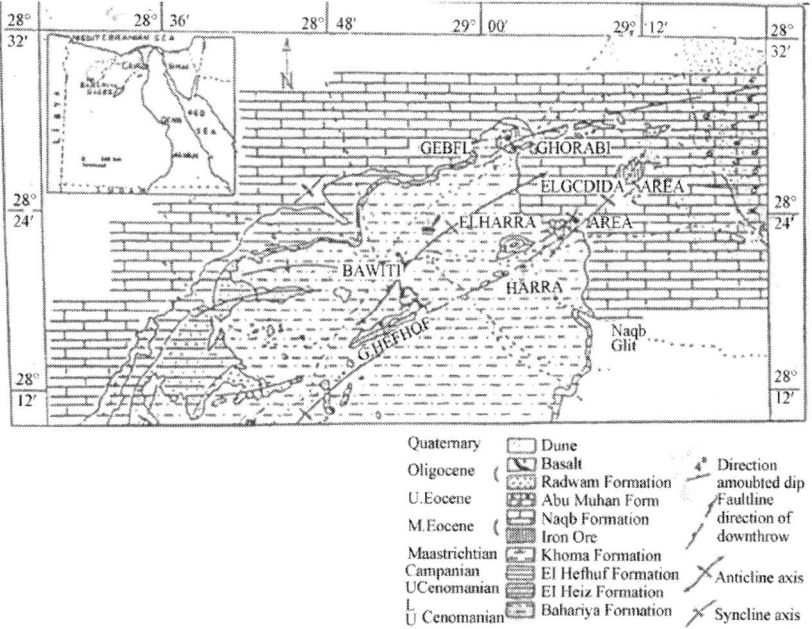

Figure 4: Geological map of El Bahareya Oases The important iron minerals localities [51].

GOLD IN EGYPT

Gold is recorded in Egypt in more than 95 occurrences most of them were mined during Pharaonic age. Fifty years ago on ward, extensive efforts were done by the Geological Survey of Egypt (EGSMA) to explore gold in the old mines areas and their vicinity. The earlier work was conducted in co-operation with Russian Experts. Through this exploration work new targets were introduced such as alteration zones around gold bearing quartz veins and banded iron formation. Also new areas outside the known old mines were explored such as South Sinai and South Western Desert [7, 38, 43, 44, 62-67

]. Several attempts were done to classify gold mineralization, among these is the early classification mentioned by Kochin and Bassyouni [28], where they classified this mineralization on the bases of the mode of occurrence and nature of mineralization into three types namely:

1-dyke type, 2-vein type, and 3-placer deposit. El Ramly et al. [68, 69] classified the gold deposits according to their geographic situations into five regions, Figure 5. Figure 5 shows the scattered occurrences of the ancient gold localities in the Eastern Desert from Latitude 28° North down to 22° North. It also shows the five geo-graphic regions according to El Ramly classification of the gold deposits in the Eastern Desert. Gold deposits were classified, based on the tectonic setting models applied to the evolution of Arabian-Nubian shield, into four main formations; goldsulphides formation, skarn gold ferruginous formations, gold sulphide Formations, and quartz vein formation. On the bases of tectonic setting proposed to the evolution of Arabian-Nubian shield, Botros [70] proposed a classification for gold mineralization as summarized in Table 2.

It is obvious from Table 2 that gold mineralization occurs in almost all the island arc and syntagmatic stage. This simply means that there is no specific lithology that could be responsible for gold mineralization, but certain gold mineralization could be hosted in certain lithology. Up till now, economic gold deposits in Egypt are related to quartz veins and adjacent alteration zones.

Table 1: Iron ores information [27]

Area	Location	Reserves, M tons	Produc.., 1000 t/y	Average Assay, Fe %	Associated constituents	Remarks
Aswan	N-E Aswan	25		44		Hematite
	S-E Aswan	300	20	46	Mn, S, P, 5102, Ti	+hydrated
	West Aswan	na		44		oxides

Eastern Desert (O3IF)*	W.Kareem	18		45		
	Um Nar	14		46		
	Abu Marwat	7		44		
	W.Dabbah	6		38		
	Um Khamees	6		44		Magnetite+
	G. Elhadid	4	-	46	Si,Ti,P,Ca	Hematite
	Abou Rakab	5.7		37		
	El-Hendousi	0.2		31		
	Urn Shaddad	0.3		48		
	Sitro	0.55		43		
Bahariva Oases	Gedida	90	3,000	53		
	Ghorabi	60		47	Hematite+	
	Nasser	30		44	MS,P,Cl,Si	Geothite+
					P,Si	Limonite+
	Al-Harrah	35		45		
	Al-Haiz	100		30	hydrogeotjite	
Uwaynat (B1F)*	G.Nazar					
		15-300 m				Magnetite+
	G.Kamel G.Arkenu	Thickness		12-42	Si,Au,Ag, Cu,Pb,Zn	Hematite+
Black Sands) Sinai		Several kms Extention				Silicates
	Rosetta-Rafah	400			Si,Ti,Zr,Th, garnet	Beach sands
		na	150	Na		

* Banded Iron Formation, BIF.

Most of the old mines have dumps and tailings containing appreciable amount of gold. This may be due to the primitive technologies existed at that time or due to sudden shut down of mines. Gold mineralization are found in lithified placers, especially along basement sediments contact zones, and disseminated gold, which probably occurs in mafic and intermediate rocks. This postulation should be tested. In this article, only one example is dis-cussed here, Barramiya gold mines.

The Barramiya gold mine represents one of the important gold mines in central Eastern Desert. It lies in mid-way between Idfu on the Nile bank and Mersa Alam on the Red Sea coast. The mineralization is confined to gold bearing quartz veins and adjacent alteration zone. The country rocks are mainly composed of ophiolitic mélange where serpentinite forms a block in the Western part which is transformed autometasomatically into talc-carbonate in the central and Eastern part. The mineralized quartz vein traverses actinolite-tremolite schists and graphite schists. Graphite schists crop out around quartz vein and may be acted as geochemical barrier and play essential role in mineralization. Granitoid rocks of calcalkaline affinity (G1 type) are intruded in North and South of the mine area. The quartz vein takes E-W strike trend with dipping Northward by high angle (75°-85°), and occupy the main axis of syncline. The average thick-ness of the quartz vein is one meter and extends for about

800 meters the alteration zone around the vein is about 6 meters thick, and mainly consists of intensively altered graphite and tremolite-actinolite schists with ferruginous, sercitization, kaolinazation process. These alterations form zonation's arranged as mentioned. Listweanite was formed as lenses and bands as a result of combination of silica and carbonate, and found to be gold bearing spots. This mine has been subjected to detailed exploration work by the Geological Survey of Egypt in co-operation with Russian experts [71]. According to this detailed study, the average gold in the quartz vein is estimated as 1.59 g/t, alteration zone 2.74 g/t, and listweanite 1.37 g/t with total reserve of 30 tons of gold [72].

Table 2: Classification of gold deposits [70]

Class	Type of deposit	Tectonic environment	Type of mineralization
Strata bound deposit	a. gold hosted in Algoma type BIF b. Au hosted in tuffaceous sediments c. Au hosted in volcanogenic massive sulphide deposits	Immature island arc Mature island arc environment	Syngenetic mineraliza-tion

Non-strata bound deposits	A. vein type mineralization	Continental margin environ-ment	Epigenetic mineraliza-tion
	1- Auriferous quartz vein hosted in metamorphic	Intraplate environment	
	rocks and / or granitic surrounded	Continental margin and intra-plate	
	2- Auriferous quartz veins in sheared		
	ophiolitic ultramafic rocks		
	3- Auriferous quartz vein as associated		
	with porphyry copper		
	4-Auriferous quartz vein at contact		
	younger gabbro-granite		
	5- Small amounts in quartz veins of		
	Sn,W,Ta,Nb mineralization		
	B. Disseminated type hosted in		
	hydrothermally altered rock		
Placer deposits	A. modern placers	Intra plate environment	
	1- Alluvial gold in wadis		
	2- Beach placers		
	B. Lithified placer		

The most important recent gold activities in Egypt today are: El Sukkary area (Centamine Limited, Pharaoh gold mine) and Hamash area (Hamash Company). Table 3 presents the published information about the two areas [71]

In both areas, El-Sukkari and Hamash, commercial production has just started. It is planned that surface mining, open pit-open cast mining, will be the mining technique. These are the most economic mining techniques for such large, low grade (average gold content is 1.5 g/t) ores.

At Hamash, heap leaching will be used for dissolving the precious metals. The heap leaching will be preceded by crushing the run-of-mine ore and screening it to pass 10 mm. This screened product will be agglomerated, using cement as a binder, and piled into heaps. The heaps will be sprayed with cyanide solution, 0.5% - 1.0 % concentration, at pH 10 - 11 for about 80 - 100 days. The pregnant solution will be passed on activated carbon to adsorb the precious elements. These elements will be stripped by hot cyanide solution. Gold winning will be carried out by electro winning, followed by gold purification to obtain gold bullion containing more than 99.5 % gold. Although the commercial production at Hamash has not started yet, pilot scale testing proved the viability of the process.

At El-Sukkari area, the crushed ore will be finely ground for agitation leaching. The pregnant cyanide solution will be treated in a manner similar to that in Hamash operation. The planned production rate at El Sukkari location is about 7 tons of gold per year.

The mineral processing plants in both locations are quite similar to the conventional worldwide techniques. They are as advanced as they can be.

Recently, gold was discovered at Uwaynat area, G. Kamel and G. Nazar. The gold assay in this area is up to 14 g/t [7, 11, 38, and 48]. This area is recently bided for exploration and exploitation [73].

Recent Bids for Gold Exploration and Exploitation

There are Bids that were decided upon in 2007 for gold ex-ploration and exploitation [73]. These bids are for the loca-tions of: Um Balad, El-Fawakhier, Fatiri, Abu Mar-wat, Wadi Kariem, Hodine, Dungash, Uwaynat, and Barramiya.

As a matter of fact, a second bid for gold exploration and exploitation has been announced to the public for additional areas in the Eastern Desert.

The Future of gold mining and processing in Egypt is a bright one. Extensive exploration, mineralogical, petro-graphic, and processing research work is necessary for profitable exploitation of the gold resources in Egypt.

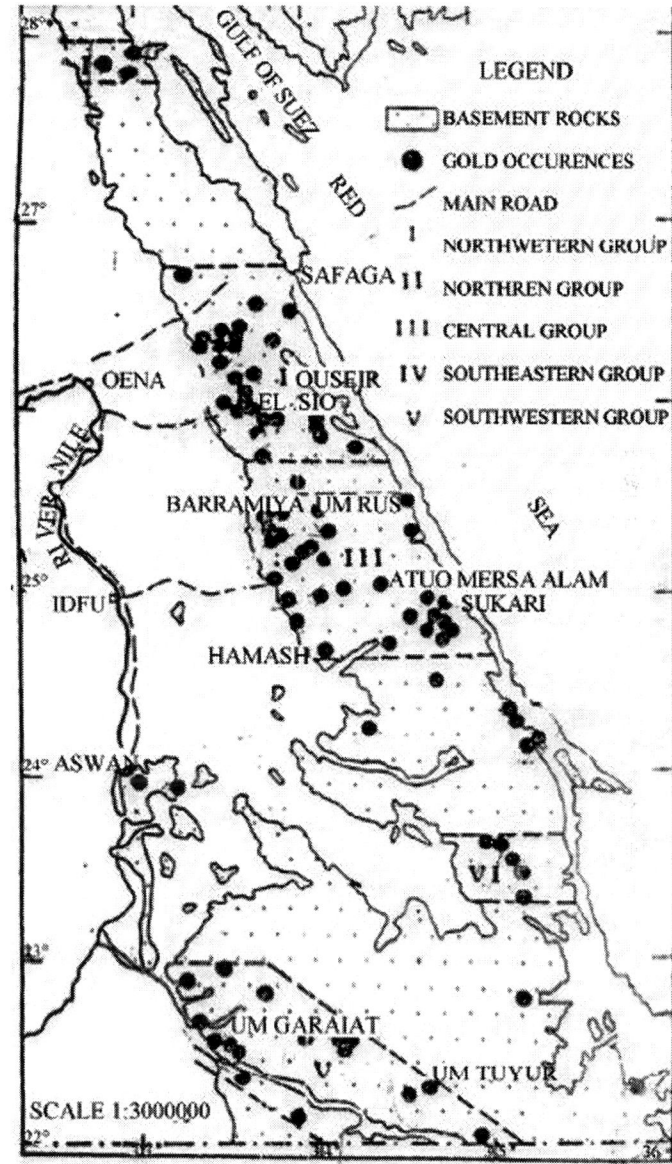

Figure 5: Gold mineralization areas as classified into five different regions [68, 69].

The history of more than 100 ancient gold workings is an indication of a huge gold source in the subsurface in the Eastern Desert.

INDUSTRIAL METAL OXIDE DEPOSITS

Tin-Tantalum-Niobium Deposits

Tin-Tantalum-Niobium deposits are genetically related to late phase of granitic intrusions which generally form small and simple intrusions of alkali to peralkali intra-plate anorogenic granites. Most of the Sn, W, Mo, Nb-Ta, REE, Be and F deposits are associated with this type. The geological, mineralogical, and geochemical characteristics of this mineralization let many authors to propose the met somatic origin [74]. The famous occurrences of this mineralization, which reach the economic level, are Abu Dabbab, Neuweibi, Muelha, Um Naggat, and Abu Rusheid. In the present article four areas are mentioned in some details.

Table 3: Recent active gold projects in Egypt. El Sukkary and Hamash gold areas information [27]

a- El Sukkary area				
Information	Grade, g/t		Total amount of rock, Mt.	Gold content, M oz.
	(Average)	1.42	64.53	2.944
Grades and quantities	(More than)	2.07	33.43	2.223
	(More than)	16.50	0.15	0.082
	(More than)	40.40	0.15	0.2000
Total at cutoff grade 0.5 g/t	(Grand average)	1.48	64.53	3.226
b- Hamash areas				
Area	Average grade, g/t	Total amount of rock, Mt.	Gold content, M oz	Recoverable gold, M oz *
Um Tondob	0.8	120	3.2	1.2
Ara	1.5	5.0	0.25	0.166
Hamash old mine	2.0-4.0	2.022	0.20225	0.022

Abou Tarda	1.5-5.0	0.34825	0.040825	0.32
Total proved		127.37	3.693075	1.708
Total probable	0.5-1.0	2,000	4.0	1.667

Abu Dabbab: It is located about 20 km North of Mersa Alam in the intersection of latitude 25°20' 27" N and longitude 34° 32' 30" E. The mineralization occupies an area of about 0.06 km2 forming cone-like shape of 100 m × 130 m. According to Sabet et al. [74], the tin, Tantalum, and Niobium mineralization is restricted to apogranite of the albitite type characterized by complex internal structure and interrelation of metasomatic facies. This granite muscovite-microcline-quartz-albite apogranite makes up the upper, central, and lower facies. The met somatic facies of the fissure zones, the greizen zone, and mineralized quartz-feldspar veins are subordinate. Mineral logically, the ore consists of albite, microcline, quartz, muscovite, and topaz. Tantalitecolumbite, pyrochlore, cassiterite, monazite, zircon, rutile, magnetite, galenite, and sphalerite are the main ore minerals.

The reserve was estimated by several authors. Sabet et al. [74] reported reserve as 20.6 million tons according to category C2 with 274 ppm Ta_2O_5, 270 ppm Nb_2O_5 and 1080 ppm Sn on the average, whereas Anonymous [75] stated that the rock reserves of Abu Dabbab is in the or-der of 40 Mt. Naim et al. [76] reviewed the Abu Dabbab ore reserves and calculated them as 7.3 million tons of ore containing 0.0266% Ta_2O_5, 0.0123% Nb_2O_5, and 0.0165% SnO_2. *Gabal Nuweibi ore Latitude 25°12› N and Longitude34° 30› E:* In this area, the ore is represented by fine dissemination of tantalite, columbite, with sub-or-dinate casseterite, fluorite, muscovite, accessory garnet, zircon, and molybdenite in apogranite rocks. The reserves are estimated by Naim et al., [76] as 114.7 million tons of low grade ore with 161 ppm Ta_2O_5, and 91 g /t Nb_2O_5 *Um Naggat:* It is located in wadi Um Gheig, South of Qusseir. It is composed of albitized and greizenized pockets in granitic rocks. Tantalite and columbite occur as disseminated minerals in the pockets at assays of 0.022% Ta_2O_5 and 0.2% Nb_2O_5.

Abu Rusheid: The ore in this area is represented by disseminations of columbite, casseterite, monazite, xeno-time, fluorite, zircon, thoragen, and microcline in mica apogranite rocks. The mineralized rock formation is the upper most part of the pssammitic gneiss that was sub-jected to metasomatic alteration. The analysis of core samples reflects that the ore contains 0.3% Nb_2O_5 and 0.033% Ta_2O_5.

Tin-Tungsten-Molybdenum Mineralization

This type of mineralization was grouped by Hussein [32] as deposits associated with granitic rocks generally of G2 and G3 types. Igla is an example of the tin-tungsten ores near Mersa Alam, Eastern Desert [77].

Molybdenum Mineralization: It occurs as disseminated and vein type at Gabal Gattar, Abu Marwa, Abu Harba, Um Disi and Homer Akarm [68, 32] Molybdenum minerals is associated with casseterite forming lateral zonation started by Molybdenum and followed by Tin minerals.

Tungsten Mineralization: Tungsten minerals occur usually in association with tin minerals in Muelha and Igla where in the latter area, Sn reaches up to 0.5 % and W up to 0.06%. Also tungston associates Sn in Abu Dabbab. There are some areas where W is the principal mineral like in Abu Hammad, and Um Bisilla [32].

Tin Mineralization: Tin occurs as disseminated and vein type in Muelha, Abu Hammad, Fatira, Abu Kharif, Abu Dabbab, and Nuweibi. At Muelha area the granite is subjected to metasomatism and the rock is albitized and altered to albite-microcline-quartz-Li-mica rock. Greisens form lenses and quartz veins with some impregnations of fluorite, cassiterite, powellite and Cu-Fe sulphides.

Tin in Placer Deposits: Tin was reported in several areas in Eastern Desert especially in the areas around the main sources. Among these areas, Igla is the famous [77] which lies west of Mersa Alam. The geological reserves were calculated by Anwar *et al*. [78] as 245000 tons as category D3 and as 170 000 tons Sn on D2 category.

Tin-tantalum-niobium ore at Abu Dabbab is a joint venture between Egypt and Gibbsland Co [79]. It is called Abu-Dabbab tantalum-tin-feldspar Project. The reserves are estimated by 40 Mt of ore. The feasibility study is based upon a design throughput of 1.26 Mt/y, which is expected to go up to 2 Mt/y. In the first stage, the production is estimated to be about 195 t/y Ta_2O_5 along with 980 t/y of tin metal during the first 20 years of production. The technical information about this joint venture is tabulated in Table 4, which presents the re-serves in the close by locations in Abu Dabbab Valley. These reserves are large enough compared with similar worldwide Tin-tantalum-niobium ores in the world. It is now in the stage of development and site preparation [80]. It is planned that the ore will be mined by a surface mining technique. The capacity of the mineral processing units will be of about 2 million tons/year. The processing plant will consist of a size reduction section (crushing and grinding using jaw crushers, and a SAG mill), flotation, and magnetic separation followed by a dewatering system to produce concentrations of tin and tantalum- niobium oxides products. The Ta-Nb Product will be used to produce Ta-Nb ferroalloys, Ta oxide, Nb oxide, and ferroniobium alloys [80]. The 40 Mt Abu Dabbab project is owned by the Egyptian registered company Tantalum Egypt, in which Gibbsland has a 50% interest by way of an incorporated joint venture with the Egyptian Government.

TITANIUM ORES

The main source of titanium in Egypt is ilmenite. Ilmen-ite is present in a rock form in different localities in the Eastern Desert, and in the black sands on the Eastern part of the Mediterranean Coast. Titanium ores information is presented in Table 5.

Ilmenite and Titaniferous Iron Ores (Rock Form)

Ilmenite and titaniferous iron ores exist in Egypt in at least 10 localities with several dimensions. They are al-ways associated with gabbroic rocks and formed by segregation. Among these areas are Abu Ghalaqa, Korab-kanci, Kolmnab, Abu Dahr, and Um Effin. The two most economically promising deposits are those located at Abu Ghalaqa and Korabkanci.

Table 4: Tin-Tantalum-Niobium ores information [79, 80]

Area	Area Location	Reserves, M tons	Produc., 1000 t/y	Average Content,
Eastern Desert	Abou Dabbab	40		0.027 % Ta_2O_5, 0.020 % Nb_2O_5, 0.017 % Sn.
	Nuweibi	115		0.017 % Ta_2O_5, 0.015 % Nb_2O_5,
	Um Naggat	25		0.022 % Ta_2O_5, 0.200 % Nb_2O_5.
	Abu Rusheid	Na		0.033 % Ta_2O_5, 0.300 % Nb_2O_5.

Abu Ghalaqa Ilmenite: This area lays 17 km South West of Abu Ghosoun port on the Red Sea coast and 100 km South of Mersa Alam city. The Ilmenite deposit is the largest among the ilmenite localities in Egypt. It is con-fined to gabbroic mass and occurs as a sheet-like body taking NW-SE and SE trend, and dips 30° to the NE direction The main ilmenite mass forms a big lens with exposed length about 300 m, and an average width of about 150 m. The detailed studies given by Hussein [32] revealed the presence of three types of ilmenite ore:

- Red ore or oxidized zone on the surface,
- Black ore or the main body, and
- Disseminated ore

The mineralogical studies showed that the ore contains the following minerals:

Ilmenite	67.4% - 68.8%
Hematite	13% - 18%
Secondary hematite	15%
Pyrite	0.13% - 2.1%
Other minerals	4% - 11%

The overall chemical analysis of the ore is:

Oxide	oxidized zone	fresh ore
TiO_2	37.09% - 41.04%	33.9% - 37.65%
Fe_2O_3	17.47% - 23.0%	6.34% - 23.85%
FeO	27.93% - 35.63%	25.94% - 31.33%
V_2O_5	0.3%1 - 0.38%	0.29% - 0.39%

Korabkanci titano-magnetite ore: This area lies in the South East corner of Egypt. According to Makhlouf et al. [81], the ore occurs as seven layers concordant with layered mafic-ultramafic assemblage. These layers are of steep exposure that dips mostly 80° - 90° to the East. The ore bands occur in parallel layers taking NNE-SSW and extend to about 2 500 m with width 50 - 80 m. The deposit exhibits medium to coarse grained texture. Mineralogically, it is composed of titan magnetite, ilmenite, hematite, goethite, sulphides with some olivine gangue. The ore could be classified into massive and disseminated ore according to the percentage of opaque minerals in the rock. The massive part of the ore contains about 80% or more of opaque minerals.

Table 5: Titanium and titaniferous iron ores [27, 41]

Area	Reserves, M tons	Produc.,, 1000 t/y	Average Assay, Ti O2 %	Associated constituents
Red Sea Coast	40	120	30-38	Fe_2O_3, SiO_2, Clays, Silicates
Mediterranean Coast(Black sands)	400		2-3	SiO_2, Mag., Rutile, Zircon, monazite

Black Sands

Black sands in Egypt are beach placers deposited from the Nile stream during flood seasons reaching the Mediterranean Sea at river mouth. It spreads on the beach East of Rashid branch of the Nile and extends east to Rafah passing through El Arish coastal plains [40]. Figure 6 shows the geographic distribution of the black sands in Egypt. They spread along the Mediterranean Sea shore from Alexandria West to Rafah East. The black sands contain some economic minerals such as ilmenite, hematite, rutile, magnetite, zircon, garnet, and monazite. Some areas were studied in details and are briefly summarized here.

Rashid East: This area is located 6 km North East of Rashid, where the area is generally flat. Heavy concentrated black sands are deposited in a thin mantle near and parallel to the shoreline. The thickness of the deposited layer ranges from 0.5 m to more than 40 m. The concentration and extension of the black sands to the West of Rashid are of negligible economic value. According to Naim *et al.* [41], the reserves of economic minerals at Rashid area are as follows (in 1000 tons):

Ilmenite	2087
Magnetite	1437
Hematite	214
Zircon	81
Rutile	29
Garnet	72
Monazite	31
Sulphides	86
Heavy silicates	1315

The ore shows lateral variations where the high concentrate occurs in the West and decreases gradually to the East.

Al Arish and Rommana Areas: These areas extend from 2 km West of Al Arish to the East of Sabkhat El Bardaweel over an area of 18 km^2. The total reserves in this area, to a depth of 1 m, are about 88 million tons with 1.1 million tons as proved ore. The proved reserves to a depth of 10 m are estimated by 3 million tons of ore. The concentration and extension of the black sands to the East of Al Arish are negligible.

The ilmenite load, at Abu Ghalaqa, is about 100 meter above the wadi level and extends to more than 200 m below the wadi level. The wadi level itself is at about 240 m above Sea Level. The major lens covers an area of 150 m × 300 m. The present status of mining technology is an open cast above the wadi level [1, 40, and 82].

The Abu Ghalaqa ore is being mined by surface mining. The benches are drilled, charged, and blasted. To facilitate loading, transporting, and crushing, secondary blasting is applied on the oversize boulders. The ore is transported to the upgrading plant nearby (about 500 m). In future it is thought to use underground mining for the lowed part of the load (below the Wadi level) to reduce the cost of overburden removal.

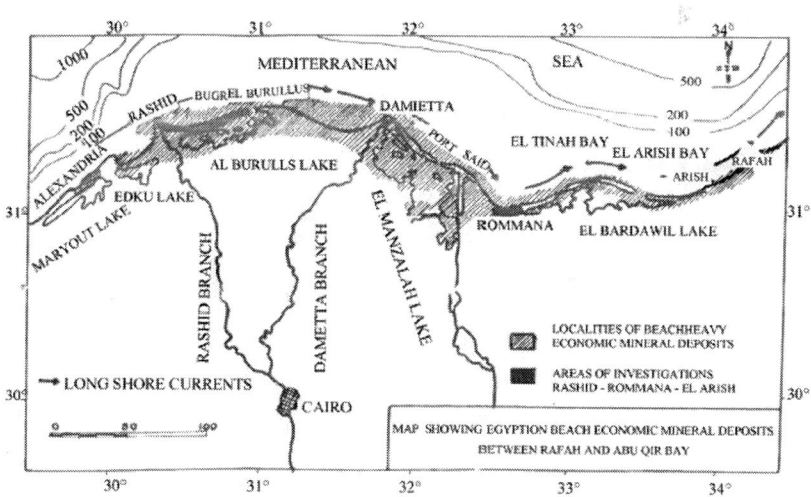

Figure 6: Locations of the Egyptian Black Sand deposits between Rashid and Rafah on the Mediterranean Sea Coast [41].

The grade of the ilmenite ore at Abou Ghalaqa is slightly upgraded by manual hand picking of some of the gangue minerals depending on difference in colors. The ore is then crushed and screened to produce different size fractions according to the end use. The only use for this ore at the time being is for coating the oil-transport pipes running under the sea water, *i.e.*, is used as heavy gravel in the concrete used for coating the oil-pipes under sea water. Laboratory experiments, up to the pilot scale, show that gravity separation, magnetic separation, and flotation produce concentrates assaying up to 43 % TiO_2. There are several researches for extracting titanium slag or titanium metal from the upgraded ore, but the results are not encouraging due to the low content of TiO_2 in the concentrate.

Black sands are dredged or scraped, piled, and trans-ported to a jungle of Humphrey spirals to scavenge out most of the green sands. The concentrate is sent to the processing plant for separating the heavy constituents [40, 41].

For the black sands, there was a plant in Alexandria for concentrating the black sands and separating its various constituents. This continued until 1970, after which the plant was shut down due to technical problems, environmental considerations as well as market saturation for the products. Nowadays, there is a pilot plant at Rosetta for developing a proper flow sheet to produce market grade products.

The main flow sheet for black sand consists of a gravity separation step to get rid of most of the green sands, followed by a low intensity wet magnetic separator to separate magnetite. The non-magnetic fraction is oven dried to be prepared for the electrostatic separation step that separates ilmenite. In a second electrostatic step, rutile is separated. After separation of rutile, the rest is taken to shaking tables to separate garnet and monazite and reject the rest of the green sands.

MANGANESE ORES

Manganese ores occur in Egypt in two major localities beside other several small occurrences [83]. The economic deposits of manganese are Um Bogma in Sinai and Elba in South Eastern Desert.

Um Bogma area lies in Central Western Sinai Manganese occurs as lenses and lensoidal bodies with different dimensions within Carboniferous sediments of Um Bogma formation. The ore shows sudden contact with dolomitic rocks and reflects stratiform type [30]. Mart and Sass [84] classified the ore into two facies dolomite facies and silty facies and they are inclined to shallow marine environmental deposition of the ore on one hand. On the other hand Attia [83] and others believed that the ore was formed during metasomatic hydrothermal process. The more recent work again supports the sedimentation origin [32].

Elba Manganese ore occurs within sedimentary rocks of Miocene age in about 24 locations in Shalateen plain.

Manganese forms vein like type striking N 120 and N 130 in a zone of 7 km width and 70 km long. El Shazly [30] proposed weathering products of basement rocks rich in manganese as a source of the ore which deposited in fissures and cracks with some replacement along frac-ture walls. Basta and Saleeb [85] suggested epigenetic origin in low temperature where manganese oxides pre-vailed and absence of silica, carbonate and sulphides which manifested in near surface deposition. Other Mn occurrences are recorded in Wadi Malik near Ras Banas and Abu Shaar El Qibli in Southern part of Esh El Mal-laha range [3, 32]. The reserves and production are given in Table 6.

Manganese is mined in the different localities by un-derground methods, mainly room and pillar. In the ex-posed outcrops it is mined by surface mining techniques. The iron oxides are highly disseminated in the manganese oxide matrix, which makes the possibility of upgrading the ore is limited.

The only processing steps carried out on the manganese ores are crushing and screening. The prepared ore is mixed with some imported high grade manganese ore and fed to the smelter at Abu Zonaima, Sinai to produce ferromanganese alloys [86]. The fines, under size fractions, are piled in dump areas. It is expected that agglomeration and magnetic roasting, followed by low intensity magnetic separation may improve the grade of the man-ganese ore in the rejected fines, which in turn may in-crease the manganese recovery.

SULFIDE MINERALIZATION IN EGYPT

Lead-Zinc Deposits

This type is located in Phanerozoic sediments and crops out in seven areas in the Southern part in the Eastern De-sert on the Red Sea Coast namely: Zug El Behar, Asel, Wizr, Um Gheig, Abu Anz, Gabal El Rusas, and Ranga [30, 36, 57, and 87]. It is restricted to lower part of Gabal El Rusas Formation which unconformable located on the Precambrian rocks in Zug El Behar and Asel areas. In the other occurrences the mineralization belongs to Upper Abu Dabbab Formation.

The primary sulphides are galena, sphalerite, and pyrite, marcasite which are transformed on the surface to ceru-site, anglesite, smithosnite, hydrozincite, jarosite, and limonite. Many authors are inclined to propose that the replacement of limegrit by cold hydrothermal solutions as the main origin of this mineralization. El Shazly [30] and El Ramly *et al.* [68] suggested synsedimentary ori-gin, whereas Hilmy *et al.* [88] pointed to the exhalative sedimentary origin. On the basis of detailed work on Um Gheig, Asel, and Zug El Behar areas, El Aref and Amstutz [89] classified the mineralization into two groups: 1) Pb-Zn as filling type in Um Gheig and Wizr, and 2) stratiform galena in Zug el Behar and Asel. The ore is restricted on the filling mass extend along the rift taking NW-SE intercontinental rift and not necessarily to be related to magmatic activity accompanying the rift. The Geological Survey of Egypt estimates the reserves in Um Gheig as 1.5 million tons with an average assay of 13.8% Zn and 2.3% Pb.

Copper Sediments

Several occurrences of copper were recorded in Phanerozoic sediments in Centre and West Sinai as secondary malachite and in some places mixed with Manganese (90). The reserves are limited.

Cu-Ni-Co Deposits

This type of mineralization is well represented in Abu Swayel and El Geneina in South Eastern Desert. The ore is closely related to mafic-ultramafic and gabbro of ophiolitic rocks.

Abu Swayel deposit is located about 185 km South Aswan. The main mineralized zone contains massive and dissimenated deposit type being hosted in amphibolite rocks, surrounded by biotite schists who could be rep-resenting the metamorphosed equivalent of ophiolitic gabbro and basalt [32]. The main minerals are pyrite, pyrrhotite, chalcopyrite, pentlandite, violarite, and limonite. Malachite occurs in the oxidized zone. The reserves are estimated as 185 000 tons containing 2.8% Cu and 1.57% Ni and minor concentration of Cobalt [32].

El Geneina deposit is located in the intersection of Latitude 23°57'N and Longitude 34°37' E where gossans with Cu and Ni do exist. Malachite and garnierite stained gossans are associated with thrust slicks of mafic rocks. Ore minerals are pyrite, pyrrohotite, and chalcopyrite. The metal assays are 0.17 % Cu and 0.38 % Ni [32].

Table 6: Manganese ores information [27]

Area	Location	Reserves, M tons	Produc., 1000 t/y	Average Content, MnO_2	Associated constituents
Eastern Desert	Esh El-Mallahah W. Ma'alik	Na	Na	35.8-45.14	Fe_2O_3, SiO_2, clays
	G.Elba & Abou Ramad	Na	120	42.17	
	W. Ma'alik	Na	Na	45.0	Fe_2O_3, SiO_2, clays
Sinai	Abou Zunima	5	120	38	

Cu-Ni Sulphide

This type of mineralization occurs in gabbro rocks at Akarm (24°00'N and 34°17'E) the mineralization exhibits both massive and disseminated types within norite, melanorite, and peridotite. The surface expression clearly shows the presence of three zones of gossans which could be related to three sulphide bands. The total re-serves of this location are estimated as 700 000 tons with 0.95 % Cu-Ni [32].

Stratiform Massive Sulphides

This type of mineralization is represented by a group of small deposits in South Eastern Desert, e.g., Um Samuki, Helgit, Maakal, Darheeb, Abu Gurdi, Egat, and Al At-shan.

Um Samuki deposit lies in the intersection of latitude 24°14' N and long 34°30 E. It is Zn-Cu-Pb deposit. The area is mainly built of Cal alkaline island arc volcanic andesite and their pyroclastics As a result of these mineralization conditions, the area received intensive studies. These studies attribute the mineralization to epigenetic process, where it was introduced by hydrothermal solutions along shear zones developed by replacement of pre-existing rocks. On the contrary of epigenetic hydrothermal deposition, Hussein *et al.* [91] and Hussein [32] believed that this deposit is a massive sulphide body which was deposited during the Abu Hamamid volcanics episode on the top of submarine volcanic vent system and the sedimentation took place conformably with the en-closing rocks at the interface between the volcanic pile and sea water. The ore bodies overlay a stock work of altered rocks resulting from intensive met somatic effects induced by the ascending volcanic exhalation on the channel ways through which they ascended [91]. The ore body in the Western part assays 2.2 % Cu, 21.6 % Zn, 0.5 % Pb and 109 g/t Ag with total reserves of 200 000 tons [92] while the Eastern part is less in metal content where Cu possess 1.8 %, Zn 13.6 % ,and Pb 3.4 %.

There are no mining activities in the sulphide mineralization areas except at Um Ghaig where the production is at very small scale. Some of these localities were exploited by the Ancient Egyptians and used in manufacturing metallic alloys such as brass and bronze. The rest of the sulfide ores in Egypt exists in small quantities, which cannot be exploited economically nowadays.

CHROMITE DEPOSITS

Chromite deposits occur as small lenses of podiform within serpentinite rocks of ophiolitic sequences at Gabal Moqassem, Um El Tiyur, Sul Hamid, Um Krush, Wadi Himur, Abu Dahr, Wadi Ghadir, Um Khariga and others. Most of these locations lie South of Latitude 26° N [32]. The majority of these ores are exhausted. The origin of chromite is attributed to early crystallization followed by crystal settling from basic magma at spreading centers during the formation of new oceanic crust which tectonically emplaced during accretion prior to cratonization [32].

No large scale exploitation is reported in any of the mentioned chromite occurrences. It is well known that chromite ores can be beneficiated by gravity separation and/or flotation depending on the ore constituents and the economic liberation size.

POSSIBLE AREAS FOR INVESTMENT IN MINERAL INDUSTRY IN EGYPT

The following areas are open for serious investment in the mineral industry, metallic commodities, in Egypt:
- Mining and Mineral Processing of iron ores at: Uwaynat (Western Desert), Eastern Desert, Baharya Oases, and Aswan.
- Integrated iron and steel industry.
- Exploitation of ilmenite ores in the feasible areas.
- Evaluation and exploitation of Beach Black Sands for their strategic hevy minerals.
- Exploration, Mining, Processing, and Extraction of: gold, tin, tantalum, and niobium.

CONCLUSIONS

The mineral resources in Egypt are plenty. However, it could be multiples of the known reserves if the appropriate subsurface exploration technology is used. Extrapolation of the available geological data suggests that with some additional geological efforts, clear ideas could be obtained about new mineral findings and/or extension of the existing deposits. As has been presented above, the simple primitive mining and mineral processing techniques limit the production capacity and produce inferior quality products, which lead to waste of resource, high cost of extraction, and low quality product.

Most of the metallic mining activities in Egypt are in the form of small operations, except for iron ore, which is reflected on the production cost being high. The most that is being done on any of the exploited commodities to upgrade or clean them is crushing, grinding, screening and sometimes grading and/or classification. Very little up-to-date technology in this area is being adopted. The concept of added value in the mineral industry in Egypt is almost missing. As a result, the low grade mineral products from such simple treatment are being marketed locally or exported. Consequently, the exported low grade mineral commodities are sold at ridiculously low prices because of reluctance to update the technology. It is recommended that large scale mining operations and processing plants, on the bases of advanced technology, are to be introduced and implemented in the mineral industry in Egypt. These will lead to improved quality, lower cost products, and more organized and inter-related mining systems.

ACKNOWLEDGEMENTS

The Authors would like to thank their colleagues who offered all kinds of help to them. Among those are Dr. A. A. Negm, Dr. M. A. El Wageeh, Dr. A. Dardir, and Mr. Wafae W. Ghobrial They provided the Authors with valuable information, and discussed the scientific mate-rial with them. Thanks are also extended to Dr. G. Oz-bayoglu and Dr. A. I. Arol from the Middle East Techni-cal University at Ankara, Turkey for suggesting the topic and inviting one of the Authors, Dr. Abouzeid, to present the content of this article at their 11th International Min-eral Processing Symposium at Belek-ANTALYA, TUR-KEY (October, 2008). Great thanks and appreciation are due to Miss. Eanass A. Abouzeid for her help in prepar-ing the Figures and putting the manuscript in its final form.

REFERENCES

1. M. A., Ghonaim, "Present and Future of the Mining Industry in Egypt," Egyptian Geological Society, Vol. 17, 1978, pp. 35-46.
2. A.-Z. M. Abouzeid, "Maximization of Added Value in Mineral Processing," 9th International Mining, Petroleum, and Metallurgical Conference, Cairo University, Cairo. February 2005, pp. 135-156.
3. R. Said, "The geology of Egypt," Tay-lor and Francis Publishers, London, 1990.
4. A. Richter, "Geologie der Metamorphen und Magmatischen Gesteine in Gebiet Zwischen Gebel Uwaynat und Gebel Kamel, SW Agyp-ten, NW Sudan," Berl. Geowiss. Agh, Vol. 73, 1986, pp. 1-201.
5. A. Richter and H. Schandelmeier, "Precambrian Basement Inliers of Western Desert, Geology, Petrology, and Structural Evolution,"

In: R. Said, Ed., Geology of Egypt, Tay-lor and Francis, 1990, pp. 185-250.
6. EGPC and CONOCO, "Egyptian General Petroleum Corporation and CONOCO," A geological map of Egypt, 1987.
7. G. Naim, A. M. Khalid, G. M. Said, S. Shaaban, A. Hussein and M. El Kady, "Banded Iron Formation Discovery at West Gabal Kamel and Its Gold Potentiality, Western Desert," Annals of the Geological Survey of Egypt, Vol. 21, 1998, pp. 303-330.
8. J. Klerkx, "Age and Metamorphic Evolution of the Basement Complex around Gabal Aluwaynat," In: M. J. Salem and M. T. Busserawil, Eds., the Geology of Libya, Academic Press, London, 1980, pp. 901-906.
9. M. Sultan, Z. El Alfy and K. Tucker, "U-Pb (Zircon) Ages from the Uweinat Area," Abstracts of centennial of EGSMA, Cairo, 1996.
10. A. O. Abu Salem, "Geology and Mineralogy of the Basement Rocks of West Gabal Kamel area," M.Sc.Thesis, Al Azhr University, 2003, p. 90.
11. M. Khattab, O. R. Greiling, A. M. Khalid, M. Said, A. Kontany, A. Abu Salem, et al. "Uwaynat Banded Iron Formation (SW Egypt) Distribution and Related Gold Miner-alization," Annals of the Geological Survey of Egypt, Vol. 25, 2002, pp. 343-364.
12. M. F. Abdel Fattah, "Petrological and Geochemical Studies on the Basement Rocks of Northeast Gabal Uwaynat Area, Western Desert, Egypt," 2005, PhD The-sis Suez Canal University, p. 245.
13. D. B. Stoeser and V. F. Camp, "Pan African micro plate accretion of the Arabian shield," Geological Society of America Bulletin, Vol. 96, 1985, pp. 817-826. doi:10.1130/0016-7606(1985)96<817:PMAOTA>2.0.CO;2
14. A. Kroner, R. Greiling, T. Reischamann, I. R. Hussein, R. Stern, S. Durr, et al, "Pan African Crustal Evolution in the Nubian Segment of Southeast Africa," In: A. Kroner A. Ed., Proterozoic Lithosphere Evolution, American Geophysical Union, Geodynamic series, Washington D.C. Vol. 17, 1987, pp. 1611-1634.
15. M. A., El Sharkawy, R., El Bayoumi. The ophio-lites of Wadi Ghadir Area, Eastern Desert, Egypt. Annal. Geol. Surv. Egypt. 1979; 9: 125-135.

16. M. G., Abdel Salam, R. J., Stern, Suture and shear zones in the Arabian-Nubian Shield. Jour. of African Earth Science. 1996; 31: 289-310.
17. A. A. Abdel Meguid, "Late Proterozoic Pan African Tectonic Evolu-tion of the Egyptian Part of the Arabian-Nubian Shield," Mid-dle East Research Centre, Ain Shams University, Vol. 6, 1992, pp. 13-28.
18. N. Sturchio, M. Sultan, P. Sylvester, R. Batiza, C. Hedge and A. A. Abdel Maguid, "Geology, Age, and Origin of Meatiq Dome: Implications for the Precambrian Stratigraphy and Tectonic Evolution of the Eastern Desert of Egypt," Fac. Earth Sc. Bull., King AbdulAziz University, Jeddah, Saudi Arabia, Vol. 6, 1983, pp. 127-143.
19. A. A. Hussein, M. Ali and M. F. El Ramly, "A Proposed New Classification of the Granites of Egypt," Journal of Volcanology and Geothermal Research, Vol. 14, 1982, pp. 187-198. doi:10.1016/0377-0273(82)90048-8
20. E. M. El Shazly, T. H. Dixon, A. E. J. Engel, A. A. Abdel Meguid and R. J. Stern, "Late Precambrian Crustal Evolution of Afro-Arabia from Oceanic arc to Craton, Egypt," The Journal of Geology, Vol. 24, No. 14, 1980, pp. 101-121.
21. B. Grothous, D. Eppler and R. Ehrlich, "Deposi-tional Environment and Structural Implications of the Ham-mamat formation, Egypt," Annal of Geological Survey of Egypt, Vol. 9, 1979, pp. 564-590.
22. M. F. El Ramly and A. A. Hussein, "The Ring Complex of the Eastern Desert of Egypt," The Journal of African Earth Sciences, Vol. 3, 1985, pp. 77-82. doi:10.1016/0899-5362(85)90024-7
23. B. Issawi, M. El Hi-nawi, M. Francis and A. Mazhar, "The Phanerozoic Geology of Egypt, A geodynamic Approach," Geological Survey of Egypt, 1999, p. 76.
24. L. L. Sloss, "Global Sea Level changes: A View from the Craton," Geological and Geophysical Investiga-tions of Continental Margins, A. A. P. G. Memorial, Vol. 29, 1979, pp. 461-467.
25. F. B. Van Houten, "Latest Jurassic-Early Cretaceous Regressive Facies," In: A. A. Craton and P. G. Bull, Eds., Northeast Africa, Vol. 64, 1980, pp. 857-867.

26. J. Ball, "Contribution to the Geography of Egypt," Geological Survey of Egypt, 1952, p. 308.
27. A.-Z. M. Abouzeid and M. A. El Wgeeh, "Mineral Industry in Egypt-State of the Art," 11th International Mineral Processing Sym-posium, Belek, Antalya, Turkey. 2008, pp. 1-27.
28. G. Ko-chine and F. A. Bassyuni, "Mineral Resources of the UAR, Part I, Metallic Minerals," Internal Report No. 18/19/68, Geological Survey of Egypt, 1968, p. 35.
29. A. S. Amin, "Geological Features of Some Mineral Deposits in Egypt," Bulletin De In-stitute du Desert, Egypt, Vol. 1, 1955, pp. 208-239.
30. E. M. El Shazly, "Classification of Egyptian Mineral Deposits," Egyptian Journal of Geology, Vol. 1, No. 1, 1957, pp. 1-20.
31. T. G. Ivanov, I. Shalaby and A. A. Hussein, "Metal-logeneic Characteristics of South Eastern Desert, Egypt," Annal of Geological Survey of Egypt, Vol. 3, 1973, pp. 139-166.
32. A. Hussein, "Mineral Deposits," In: R. Said, Ed., The Geology of Egypt, Taylor and Francis Publishers, London, 1990, pp. 511-566.
33. M. S. Garson and I. Shalaby, "Pre-cambrian Lower Paleozoic Plate Tectonics and Metallogenesis in Red Sea Region," The Geological Association of Canada, Special Issue, 1976, pp. 537-596.
34. N. S. Botros and A. M. Noor, "Mineral Deposits in the Eastern Desert of Egypt, an Expression of two Major Episodes with Distinct Magmatic and Tectonic Characteristics, Annal of Geological Survey of Egypt, Vol. 30, 2008, pp. 249-274.
35. W. W. Ghobrial, "Iron Ores in Egypt," Personal Contact, 2008.
36. W. W., Ghobrial, "Lead and zinc in Egypt," Personal Contact, 2008.
37. W. W. Ghobrial, "Developing of Mineral Resources in Egypt," Per-sonal Contact, 2008.
38. A. M. Khalid and A. A. Diaf, "Geo-logical and Geochemical Exploration for Gold and REE at Ja-bal Nazar and Jabal Arkenu, Egypt-Libya," Proceedings of Geological Survey, Egypt, 1996, pp. 425-446.
39. A. M. Khalid, O. R. Greiling, M. M. Said, A. Megahed, G. Shaaban, M. Micheal, et al., "South Western Desert BIF Laboratory Studies

and Gold Extraction Tests," Annal of Geological Sur-vey of Egypt, Vol. 25, 2002, pp. 315-332.

40. M. A. Hassan, "Black Sands Project," A Briefing to the Egyptian Association for Mining and Petroleum, Nuclear Material Authority, Cairo. June 12 2003, p. 21.

41. G. Naim, E. T. El Melegy and A. El Azab, "Black Sand Assessment," The Egyptian Geological Survey, 1993, p. 67.

42. Hunting Geophysical Co. Geology of Al Uwaynat, East. Libya, IRC Tripoli. Internal Report, 1974, p. 190.

43. N. Sh. Botros, "Geological and Geochemical Studies on Some Gold Occurrences in the North Eastern Desert," Ph.D. Thesis, Zagazig University, Zagazig, Egypt 1991, p. 146.

44. A. Dardir and K. El Chimi, "Geology and Geo-chemical Exploration for Gold in the Banded Iron Formation of Um Nar Area, Central Eastern Desert, Egypt," Annal of Geo-logical Survey of Egypt, Vol. 18, 1992, pp. 103-111.

45. M. F. El Ramly, M. K. Akaad and A. H. Rasmy, "Geology and Struc-ture of Um Nar Iron Deposit," Special Paper, No. 28, Geologi-cal Survey .Egypt, 1963, p. 29.

46. P. K. Sims and H. James, "Banded Iron Formation of Late Proterozoic Age in the Central Eastern Desert of Egypt, Geology and Tectonic Setting," Eco-nomic Geology, Vol. 79, 1984, pp. 1777-1784. doi:10.2113/gsecongeo.79.8.1777

47. A. El Dougdoug, M. F. Awadallah and Z. Hamimi, Textural Relations in the Banded Iron Formation Facies of Gebel El Hadid Area, Central Eastern Desert, Egypt," Annal of Geological Survey of Egypt, Vol. 15, 1985, pp. 31-44.

48. M. Said, A. M. Khalid, M. El Kady, A. Abu Salem and S. Ibrahim, "On the Structural Evolution of Banded Iron Formation of Gabal Kamel and Its Role in the Gold Mineralization," Annal of Geological Survey of Egypt, Vol. 21, 1998, pp. 345-352.

49. D. D. Klemm, "The Forma-tion of Paleoproterozoic Banded Iron Formation and Their As-sociate Fe and Mn Deposits with Reference to Griqualand West Deposits, South Africa," The Journal of African Earth Sci-ences, Vol. 30, No. 1, 2000, pp. 1-24. doi:10.1016/S0899-5362(00)00005-1

50. E. Basta and H. Amer, "El Gidida Iron Ores and Their Origin, Bahariya Oases, Egypt," Economic Geology, Vol. 64, 1969, pp. 424-444. doi:10.2113/gsecongeo.64.4.424
51. A. A. El Bassyony, "Ge-ology of the Area between Gara El Hamra, Ghard El Moharik and El Harra Area, Bahariya Oases, Egypt," M. Sc. Thesis, Cairo University. 1970, p. 98.
52. A. A. El Bassyony, "Geo-logical Setting and Origin of El Harra Iron Ores, Bahariya Oa-ses, Western Desert, Egypt," Annal of Geological Survey of Egypt, Vol. 23, 2000, pp. 213-222.
53. S. Akaad and B. Is-sawi, "Geology and Iron Deposits of Bahayria Oasis," The Egyptian Geological Survey, No. 18, 1963, p. 300.
54. M. A. El Sharkawy, M. A. Higazi and M. A. Khalil, "Three Probable Genetic Types of Iron Ore at El Gadida Mine, Western Desert," Egyptian Journal of Geology, Vol. 31, 1987, pp. 1-2.
55. M. M. El Aref and Z. Lotfi, "Genetic Karst Significance of the Iron Ore Deposits of El Bahariya Oases, Western Desert," Annal of Geological Survey of Egypt, Vol. 15, 1985, pp. 1-30.
56. M. A. Khalil, "Geological and Mineralogical Studies on the North-eastern Part of El Bahariya Oases, Western Desert, Egypt," Ph. D. Thesis. Al Azhar University 1995, p. 237.
57. E. M. El Shazly and A. A. Hassan, "The Results of Drilling in the Iron Ore Deposit of Ghorabi, Bahariya Oases, Western Desert," Survey Depart, 1962, p. 41..
58. D. Neev, K. J. Hall and M. J. Saul, "The Pelasium Megashear System across Africa and As-sociated Lineament Swarms," Journal of Geophysical Re-search, Vol. 87, No. B2, 1982, pp. 1015-1030. doi:10.1029/JB087iB02p01015
59. B. Issawi, "Geology of the South Western Desert of Egypt," Annal of Geological Sur-vey of Egypt, Vol. 11, 1981, pp. 57-66.
60. M. I. Attia, "To-pography, Geology, and Iron Ore of the District East of As-wan," The Egyptian Geological Survey, 1955, p. 262.
61. EISC, Egyptian Iron and Steel Co. Projected Plan for 2007/2008. 2007, p. 39.
62. M. A. Khalid, M. M. Said, A. El Naggar and N. Moselhy, "Geological and Geochemical Explo-ration at Gabal Kulyeit and Its Environs, South Eastern Desert, Egypt," Annal of Geological Survey of Egypt, Vol. 23, 2000, pp. 223-233.

63. Kh. Oweiss and A. M. Khalid, "Geochemi-cal Prospecting at Um Qareiyat Gold Deposit, South Eastern Desert, Egypt," Annal of Geological Survey of Egypt, Vol. 17, 1991, pp. 145-151.
64. A. M. Khalid and Kh. Oweiss, "Re-sults of Mineral Exploration Programs in South Eastern Sinai, Egypt," Annal of Geological Survey of Egypt, Vol. 20, 1995, pp. 207-220.
65. A. M. Khalid and Kh. Oweiss, "Geochemi-cal Exploration for Gold at Wadi Kid Area, Southern Sinai, Egypt," Annal of Geological Survey of Egypt, Vol. 20, 1995, pp. 333-342.
66. B. B. Nasr, M. S. Masoud, H. El Sherbini and A. Makhlouf, "Some New Occurrences of Gold Minerali-zation, Eastern Desert, Egypt," Annal of Geological Survey of Egypt, Vol. 21, 1998, pp. 331-344.
67. I. Khalaf and Kh. Oweiss, "Gold Prospection in the Environs of Sukkari Gold Mine, Central Eastern Desert," Annal of Geological Survey of Egypt, Vol. 23, 1993, pp. 223-233.
68. M. F. El Ramly, S. S. Ivanov and G. G. Kochin, "Studies on Some Mineral Deposits of Egypt, Part I, Section A, Article 3, Tin-Tungsten Mineraliza-tion, Eastern Desert, Egypt," The Egyptian Geological Survey, 1970, pp. 120-145.
69. M. F. El Ramly, S. S. Ivanov and G. G. Kochin, "The Occurrence of Gold in the Eastern Desert of Egypt," In: O. Moharm et al. Eds., Studies on Some Mineral Deposits of Egypt, The Egyptian Geological Survey, 1970, pp. 53-64.
70. N. Sh. Botros, "A New Classification of the Gold Deposits of Egypt," The Journal of African Earth Sciences, Ore Geology Review, Vol. 4, No. 2, 2004, pp. 1-35.
71. EGSMA, "Egyptian Geological Survey and Mining Authority, Egypt," Results of Prospecting and Provisory Work for Gold at Bar-ramiya, Sukkary, Um Nar Prospects, Internal Report No. 19/77, 1977.
72. EGSMA, "Egyptian Survey and Mining Authority, Egypt," Results of Prospecting and Evaluation Carried out at the Eastern Flank of the Barramiya Gold Ore Deposit in 1976-1977, Internal Report No. 16/78, 1978.
73. EMRA, "Egyptian Mineral Resources Authority," Results of 2006 1st Interna-tional Bid Round for Gold Exploration and Exploitation in Egypt, Egypt, 2006.

74. A. H. Sabet, V. B. Tsogoev, L. M. Baburin, A. Riad, A. Zakhari and L. Armanious, "Geologic Structure and Laws of Localization of Tantlum Mineralization at Neweibi Deposit," Annal of Geological Survey of Egypt, Vol. 6, 1976, pp. 119-156.
75. Anonymous, "Egyptian Tin and Tantalum," Mining Magazine, October 2004.
76. G. M. Naim, A. T. El Melegy and Kh. Soliman, "Tantalum-Niobium-Tin Mineralization in Central Eastern Desert, Egypt, a Re-view," Proceedings of Geological Survey, 1996, pp. 599-622.
77. H. Sadek, "Report on Igla Tin Deposit, Eastern De-sert Mines and Quarries Depart," Internal Report, Cairo, Egypt, 1944.
78. T. Anwar, M. A. Morsy, A. I. Arslan and M. A. El Maky, "Estimation of the Probable Geological Reserve of Tin in the Igla Area, Eastern Desert, Egypt," Arab Mining and Petroleum Association Conference, Ismalia, Egypt, 1983, p. 39.
79. Gippsland Limited, "Tantalum Industry Overview," 2005, Annual Report, p. 58.
80. Gippsland Limited, Annual Report, 2005, p. 55.
81. A. Makhlouf, N. Y. Beniamin, M. M. Mansour, S. A. Mansour and H. El Sherbini, "Mafic-Ultra Ma-fic Intrusion of South Korabkanci Area with Emphasis on Ti-tanomagnetite Ores, South Eastern Desert, Egypt," Annal of Geological Survey of Egypt, Vol. 30, 2008, pp. 1-20.
82. El Nasr Mining Co. Monetary Status for Plan for 2006/2007, 2007.
83. M. I. Attia, "Manganese deposits of Egypt," The 20th International Geological Congress, Mexico, 1956, pp. 143-171.
84. S. Mart and E. Sass, "Geology and Origin of Manganese Ore of Um Bogma, Sinai," Economic Geology, Vol. 67, 1972, pp. 145-155. doi:10.2113/gsecongeo.67.2.145
85. E. Z. Basta and G. S. Saleeb, "Elba Manganese Ore and Their Origin, South Eastern Desert, Egypt," Mineralogical Magazine, Vol. 38, 1971, pp. 235-244. doi:10.1180/minmag.1971.038.294.13
86. ESAC, "Egyptian Steel Alloys Co. Projected Plan for 2007-2008," 2007, pp. 1-46.

87. E. M. El Shazly, A. Mansour, M. S. Afia and M. G. Ghobrial, "Miocene Lead and Zinc Deposits in Egypt," 20th International Geological Congress, Mexico, 1959, pp. 119-134.
88. M. E. Hilmy, F. M. Nakhla and M. Ramsy, "Contri-bution to the Mineralogy, Geochemistry, and Genesis of the Miocene Pb-Zn Deposits in Egypt," Chemie der Erde, Vol. 31, 1972, pp. 373-390.
89. M. M. El Aref and C. C. Amstutz, "Lead-Zinc Deposits along the Red Sea Coast of Egypt," Monograph Series on Mineral Deposits, Gebruder Bornt Rae-ger, Stutgart, No. 21, 1983, p. 103.
90. M. E. Hilmy and M. Mohsen, "Secondary Copper Minerals from West Central Si-nai," Egyptian Journal of Geology, Vol. 9, 1965, pp. 1-12.
91. A. A. Hussein, I. M. Shalaby, M. A. Gad and A. Rasmy, "On the origin of Zn-Cu-Pb Deposits at Um Smuki, Eastern Desert, Egypt," 15th Annual Meeting, Geological Soci-ety, 1977.
92. D. L. Searle, G. S. Carter and I. M. Shalaby, "Mineral Exploration at Um Samuki," U. N. Technical Report No. 36-76, Documentation Centre of Egyptian Geological Sur-vey, Egypt, 1976.

Chapter 12

Study on Operating Characteristics of Power Plant with Dry and Wet Cooling Systems

Tao Tang[1], Jian-qun Xu[1], Sheng-xiang Jin[2], and Hong-qi Wei[1]

[1]School of Energy and Environment, Southeast University, Nanjing, China
[2]Beijing Energy Company Limited, Beijing, China

ABSTRACT

The represent paper will study the performance of the power plant with the combination of dry and wet cooling systems in different operating conditions. A thermodynamic performance analysis of the steam cycle system was performed by means of a program code dedicated to power plant modeling in design operating condition. Then the off-design behav-ior was studied by varying not only the ambient temperature and relative humidity but also several parameters connected to the cooling performance, like the exhaust steam flow rate, the air cooling fan load and the number of operat-ing cooling water pumps and cooling towers. The result is an optimum set of variables allowing the dry and wet cooling system be regulated in such a way that the maximum power is achieved and low water consumption.

INTRODUCTION

There are three ways of thermal power plants' cooling systems: dry cooling system, wet cooling system, and dry and wet cooling system. In China, the wet cooling system use in the power plans commonly, but in the northwest and northeast China where the water is shortage use air cooling system. The wet cooling systems have high thermal economy, but with high water consumption. The air cooling systems can save a lot of water, but the ex-haust steam pressure is high and varying all the time for the impact of ambient temperature [1]. The wet and dry cooling systems combines the both advantages, it not only make full power when the ambient temperature is high but also with low water consumption [2,3].

The present study was inspired by the operation of a power plant with the combined wet and dry cooling sys-tem (Figure 1), placed in Northwest China. The wet and dry cooling system is composed of an air cooled con-denser in parallel with a water cooled condenser. In the wet cooling system, the cooling water which shared by two 300 MW Units, taken from the condenser passes through four wet mechanical draft cooling towers and returns to the condenser by two cooling water pumps.

The off-design performance of an air cooling con-denser or water cooling condenser separately is well deeply investigated [4-6], but the study on the perform-ance of complies wet and dry cooling system is rarely find. So the critical element of this study is the wet me-chanical draft tower. The heat transfer in cooling tower is a very complex phenomenon. But it could be described by several equations [7-9] with some simplifying as-sumptions.

The purpose of the present paper is to explore the im-pact of a dry and wet cooling system on the thermo-dy-namic performance of a power plant. This paper offers an original contribution for cooling system performance analysis by considering the dry and wet system together.

MATHEMATICAL MODEL OF DIRECT AIR COOLING SYSTEM

The Pressure of Air Condenser

Using η-NTU method to calculate the condensate tem-perature of air condenser [1]:

$$t_{s1} = \frac{D_c(h_c - h_c')}{S_{yf} v_{yf} \rho_a c_p} \frac{1}{1-e^{-NTU}} + t_{a1} \tag{1}$$

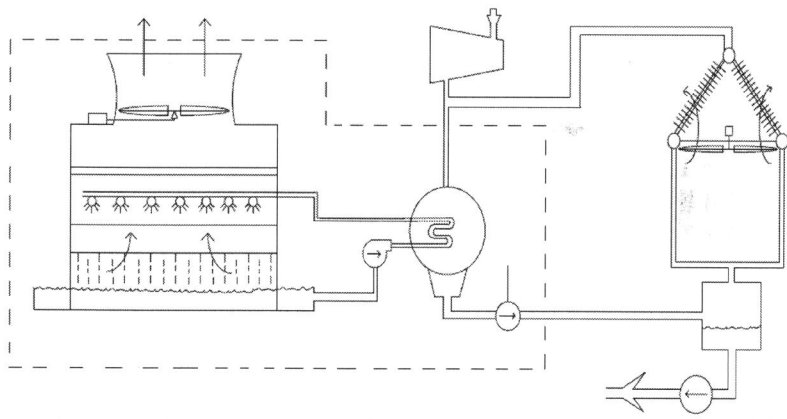

Figure 1: Schematic of the wet and dry cooling systems.

$$NTU = \frac{KS}{1000 S_{yf} v_{yf} \rho_a c_p} \tag{2}$$

where: D_c is exhaust steam flow rate. h_c' is condensate enthalpy. h_c is exhaust steam enthalpy. t_{a1} is ambient temperature. cp is air specific heat. S_y is frontal area. v_{yf} is face velocity. $_a$ is air density. NTU is heat transfer units. K is heat transfer coefficient. S is total area.

Using Equation (3) to calculate the condenser pressure of air condenser:

$$p_{s1} = 9.81 \times (\frac{t_{s1} + 100}{57.66})^{7.46} \quad (3)$$

Exhaust steam pressure is:

$$p_c = p_{s1} + \Delta p_1 \quad (4)$$

where: Δp_1 is air condenser pressure drop.

Air Cooling Condenser Heat transfer Coefficient

The total heat transfer resistance including internal ther-mal resistance, external thermal resistance, and wall thermal resistance:

$$\frac{1}{KS} = (\frac{1}{\alpha_i} + \varepsilon_i)\frac{1}{S_i} + \frac{\delta_b}{\lambda_b}\frac{1}{S_m} + (\frac{1}{\alpha_0} + \varepsilon_0)\frac{1}{\eta_0 S_{wai}} \quad (5)$$

$$S_m = (S_0 - S_i) / \ln(S_0 / S_i) \quad (6)$$

$$\eta_0 = (S_0 + \eta_f S_{chi}) / S_{wai} \quad (7)$$

where: K is total heat transfer coefficient. S is total area. α_i and α_0 are internal and external tube convective heat transfer coefficient. ε_i and $_0$ are internal and external tube fouling resistance. δ_b is base pipe wall thickness. λ_b is base pipe wall thermal conductivity. S_i and S_0 are in-ternal and external surface area of tubes. S_m is the num-ber heat transfer area of the base pipe. S_{chi} is fin surface area. S_{wai} is outer heat transfer area. η_f is fin efficiency. η_0 is the total tube fin efficiency.

MATHEMATICAL MODEL OF WATER COOLING SYSTEM

The Pressure of Water Condenser

Water condenser temperature could be got by Equation (8):

$$t_{s2} = t_{w1} + \frac{h_c - h_c'}{4.187m}[1 + \frac{1}{e^{\frac{AK}{4187Dw}} - 1}] \quad (8)$$

where $m = D_w/D_c$ is circulation ratio. D_c is steam flow rate. D_w is cooling water flow rate. h_c-h_c' is 1kg steam's latent heat. t_{w1} is cooling water temperature to condenser. K is heat transfer coefficient. A is cooling area.

Then the water condenser pressure p_{s2} can be calculated by Equation (3), the exhaust steam pressure is:

$$p_c = p_{s2} + \Delta p_2 \quad (9)$$

where: Δp_2 is water condenser pressure drop.

Cooling Water Temperature to Water Condenser

In a closed-loop cooling water system, cooling water temperature to condenser equals cooling water tempera-ture from cooling tower, it is not only affected by envi-ronmental conditions, but also by the design parameters and operating conditions of the cooling tower.

At present, the cooling tower thermodynamic calcula-tion use enthalpy method commonly [7,8]. The equations are not shown in this paper.

RESULTS AND DISCUSSION

The power plant with dry and wet cooling systems can operate as three cases: direct air cooling (Dry), dry and wet cooling system with one cooling water pump and two wet mechanical draft towers (D&W$_1$), and dry and wet cooling system with two cooling water pumps and four wet mechanical draft towers (D&W$_2$). Assuming two Units at the same operating conditions, each Unit can get half of the circulating cooling water flow rate.

In order to optimize the operation, it must study the respective operating conditions off-design characteristics fistly.

Design Parameters

Table 1 shows the main design parameters of the Unit with wet and dry cooling systems.

Each considered variable is subjected to the constraints listed in Table 2, it contains any possible the power plant operating condition during the year.

Table 1: Design parameters

Name	Unit	Content
Ambient temperature	°C	23.6
Atmospheric pressure	kPa	90.06
Relative humidity	%	86.84
Gross power output	MW	300
Exhaust steam flow rate	t/h	614.23
Exhaust steam enthalpy	KJ/kg	2437.9
Exhaust steam pressure	kPa	15
AC cooling area	m^2	492 810
AC frontal area	m^2	5128
AC face velocity	m/s	2.91
Wet condenser cooling area	m^2	3700
Cooling water flow rate	t/h	12100

| Gas-water ratio | / | 0.506 |
| CT cooling number | / | 1.12 |

Dry Configuration

The exhaust steam flow rate and ambient temperature influence on exhaust pressure can be got by using ma-thematical model of direct air cooling systems mentioned before. Figures 2-5 reports some of the parametric anal-ysis results for the Dry configuration. As can be seen from these Figs, the exhaust pressure rise with exhaust steam flow rate increases and ambient temperature rises. And the higher the ambient temperature, this trend is more obvious. The exhaust pressure will drop when AC fan load increases. The AC fan load should be operated according to the maximum load to get the highest power production, because auxiliary power consumption will rise as AC fan load rising.

Table 2: Parametric analysis ranges

Input variables	Unit	Ranges
Ambient temperature	°C	-20 - 35
Relative humidity	%	20 - 100
Air condenser fan load	%	10 - 100
Exhaust steam flow rate	t/h	250 - 650

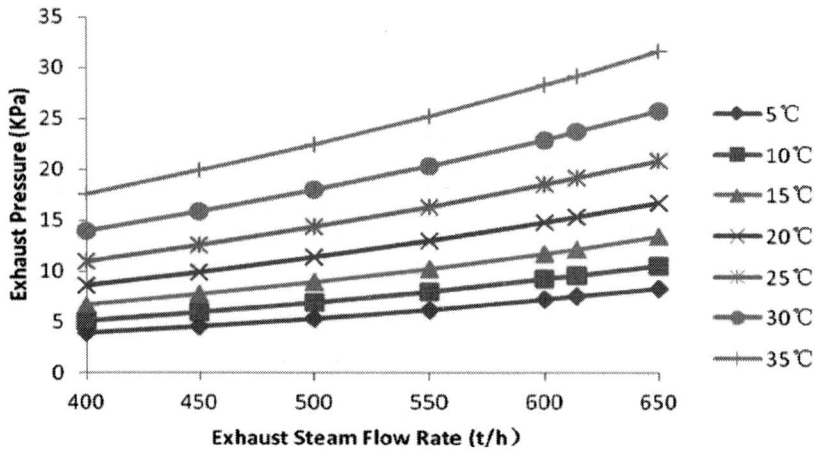

Figure 2: Exhaust steam flow rate and ambient temperature influence on exhaust pressure (AC fan load 100%)-Dry.

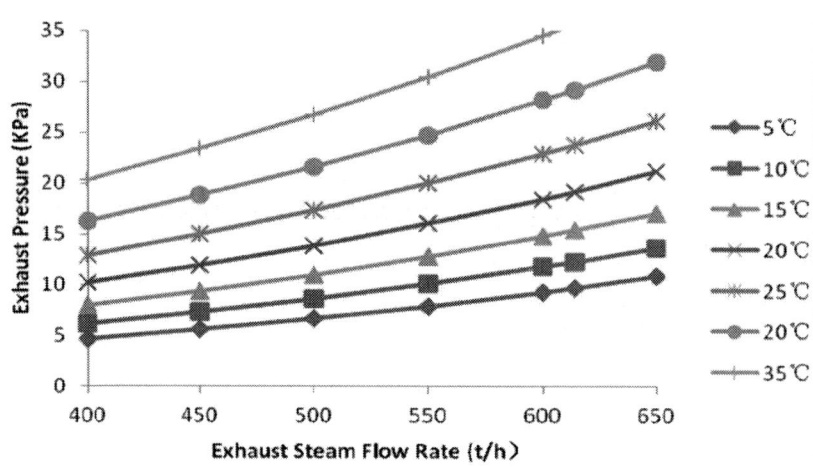

Figure 3: Exhaust steam flow rate and ambient temperature influence on exhaust pressure (AC fan load 75%)-Dry.

Study on Operating Characteristics of Power Plant with Dry ... 277

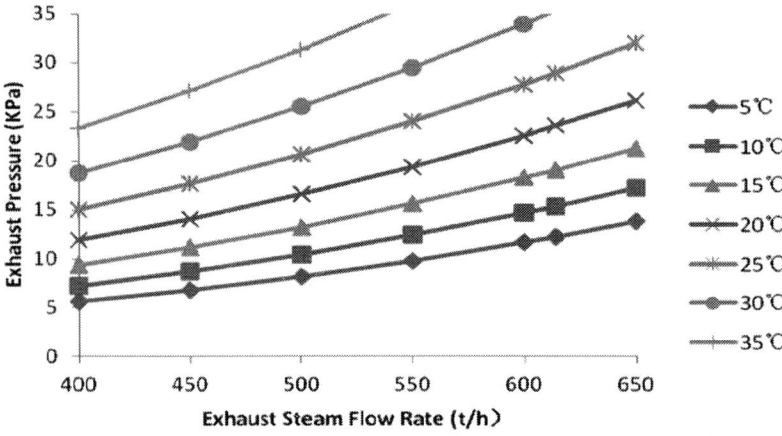

Figure 4: Exhaust steam flow rate and ambient temperature influence on exhaust pressure (AC fan load 50%)-Dry.

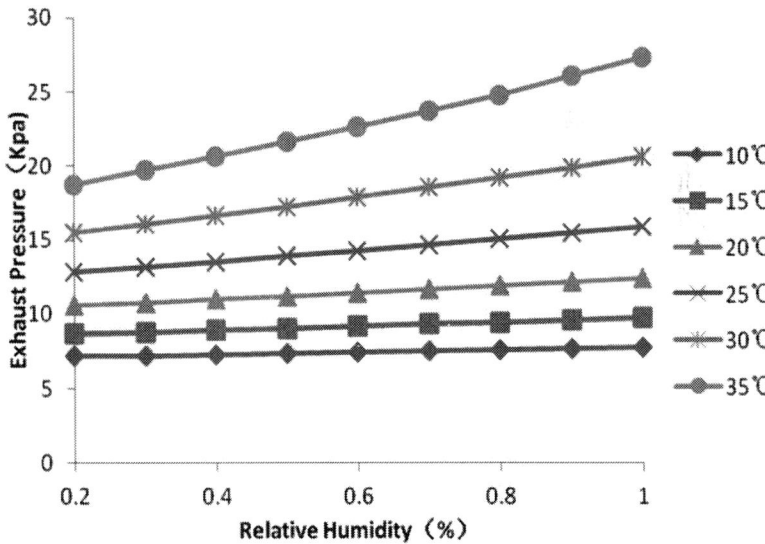

Figure 5: Relative humidity and ambient temperature in-fluence on exhaust pressure-W&D.

W & D Configuration

The wet and dry cooling system off-design process was carried out by excel VBA code by using the mathemati-cal model mentioned before.

Table 3 compares the model results against experi-mental date in three cases. The three cases show three different environmental conditions and power output. For all the cases, a good agreement was found between mod-el results and experimental date: the maximum difference is lower than 5%.

Figures 5-10 show the influence of the parameters mentioned before on exhaust pressure and dry proportion from W&D model.

Figure 5 shows the exhaust pressure variation versus ambient temperature and relative humidity. As expected, exhaust pressure increases with rising relative humidity. The effect becomes more and more appreciable with in-creasing temperature. In the same range of ambient tem-perature changes, the greater the relative humidity the more obvious exhaust pressure changes. This is consis-tent with the dry proportion behabious shown in Figure 6:

it is obvious that the steam flow rate entering the AC increases with raise in relative humidity. Relative humid-ity increases, the heat transfer capacity of the cooling tower decline, so wet proportion decreases.

The influence of the AC fan load on the cycle performance is shown in Figures 7-8. Obviously, the ex-haust pressure decreases with rising AC fan load. The effect becomes more and more appreciable with increas-ing temperature. The steam flow rate entering the AC increases with raise in AC fan load.

Table 3: Comparison between model result (M) and experimental date (EXP)

	Case 1			Case 2			Case 3		
	M	EXP	Error	M	EXP	Error	M	EXP	Error
Ambient temperature(t)	22	22	-	30	30	-	26	26	-
Relative humidity (%)	68	68	-	48.3	48.3	-	74.5	74.5	-
Power output (MW)	300	300.008	-	265	265.102	-	197	196.8	-
Exhaust steam flow rate (t/h)	614.23	-	-	540	-	-	397.5	-	-
steam flow rate to wet (t/h)	273.216	274.969	0.64%	255.11	258.594	1.37%	190.75	185.258	2.88%
Air condenser fan load (%)	93	93	-	93	93	-	94	94	-
Cooling water temperature from CT(°C)	29.287	29.768	1.64%	34.3	34.271	0.08%	30.85	31.536	2.22%
Cooling water temperature to CT (°C)	38.394	38.547	0.40%	42.804	43.584	1.82%	36.875	37.747	2.36%
Exhaust steam Pressure KPa	14.707	14.717	0.07%	17.364	17.371	0.04%	10.73	11.058	3.06%

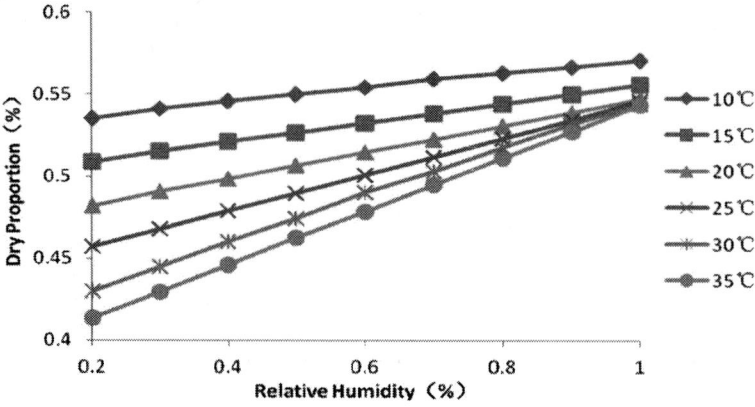

Figure 6: Relative humidity and ambient temperature in-fluence on dry proportion-W&D.

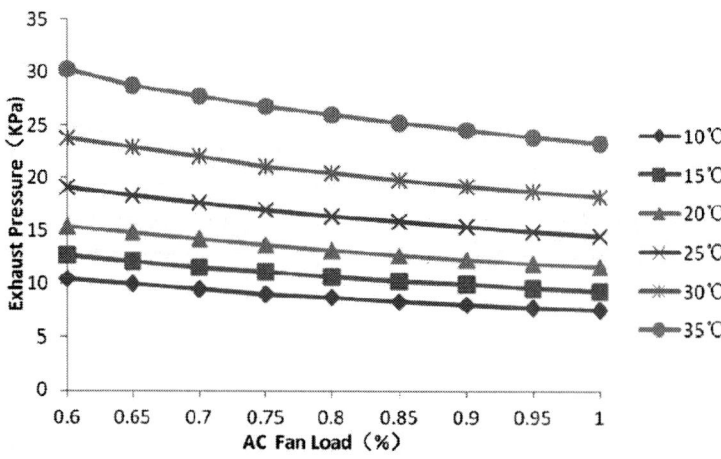

Figure 7: AC fan load and ambient temperature influence on exhaust pressure -W&D.

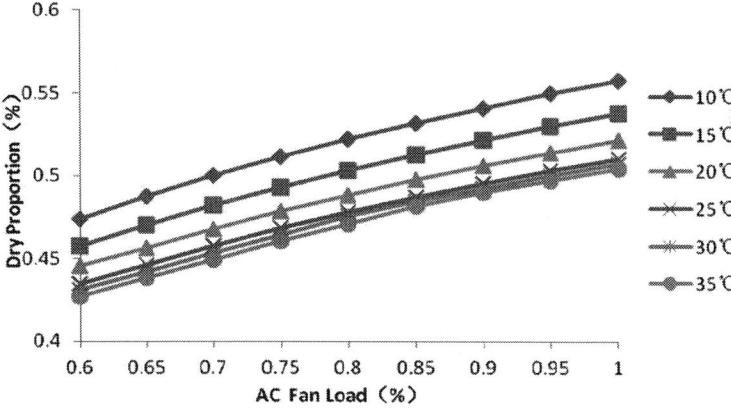

Figure 8: AC fan load and ambient temperature influence on dry proportion -W&D.

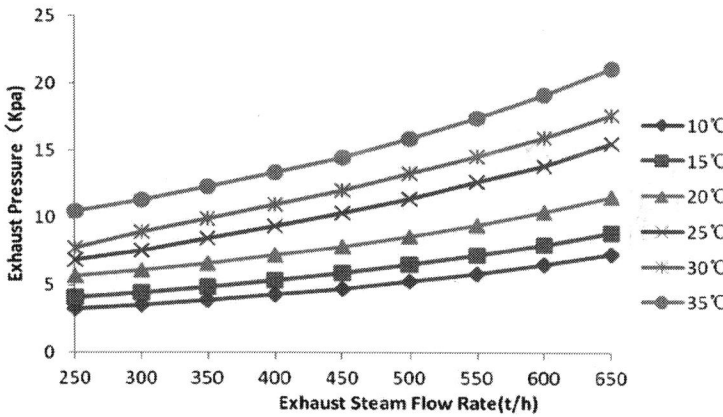

Figure 9: Exhaust steam flow rate and ambient temperature influence on exhaust pressure -W&D.

The influence of the exhaust steam flow rate on the cycle performance is shown in Figures 9-10. The ex-haust pressure progressively increases with exhaust steam flow rate increases. The effect becomes more and more appreciable with increasing temperature. The way in which the exhaust steam is shared into the two con-densers is shown in Figure 10, that a decreasing steam flow rate goes through the air cooling system as the am-bient temperature increases from 10°C to 25°C, but an increasing team flow rate goes through the air cooling system as the ambient temperature increases from 30°C to 35°C. This behavious is consistent with the water cooling condenser has better performances in high tem-perature.

Optimization of the Operation

At the low temperature, the unit operate with direct air cooling system. Figure 11 report the results of the con-densing system optimization procedure for the Dry con-figuration. The AC fan load which meet the power load at diffient ambient temperature is given in Figure 11. The AC fan load increses with rising ambint temperature. When the power load is 300MW, it must open D&W1 at 16°C, and D&W1 open at 24°C when power load is 225 MW, and D&W1 open at 31°C when power load is 150 MW. When power load is lower than 120 MW, it's no need to open D&W1.

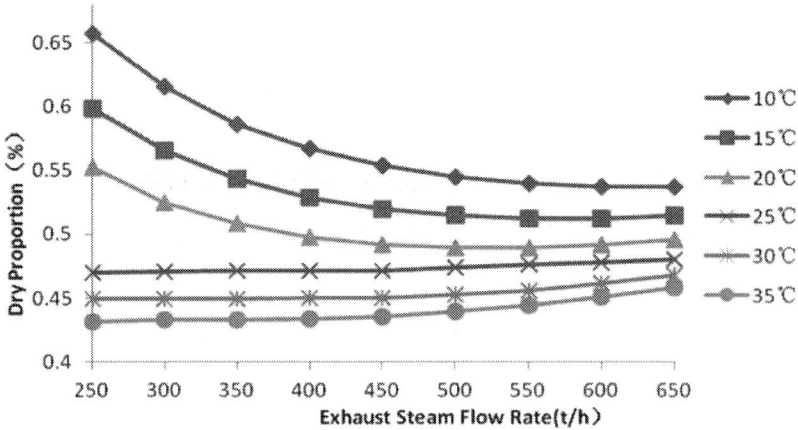

Figure 10: Exhaust steam flow rate and ambient tempera-ture influence on dry proportion -W&D.

Study on Operating Characteristics of Power Plant with Dry ... 283

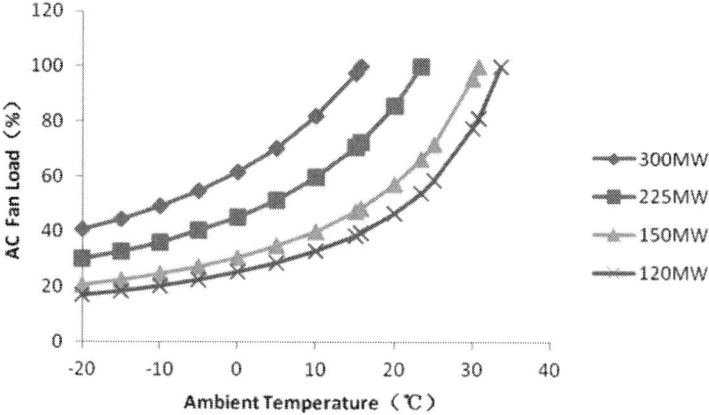

Figure 11: Power load and ambient temperature influence on AC fan load — Dry.

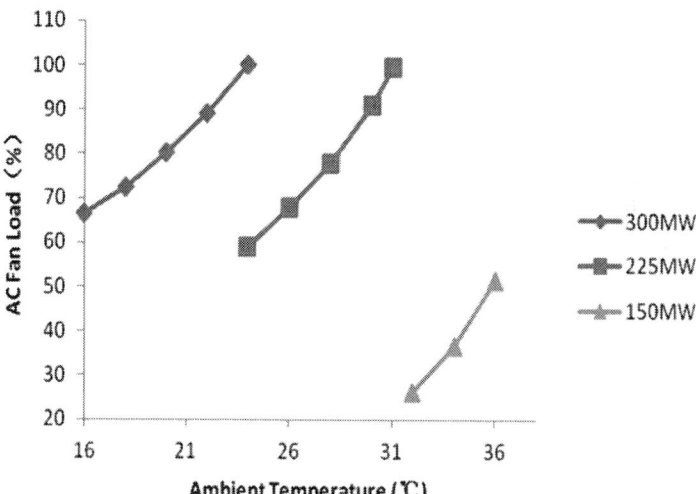

Figure 12: Power load and ambient temperature influence on AC fan load — D&W1.

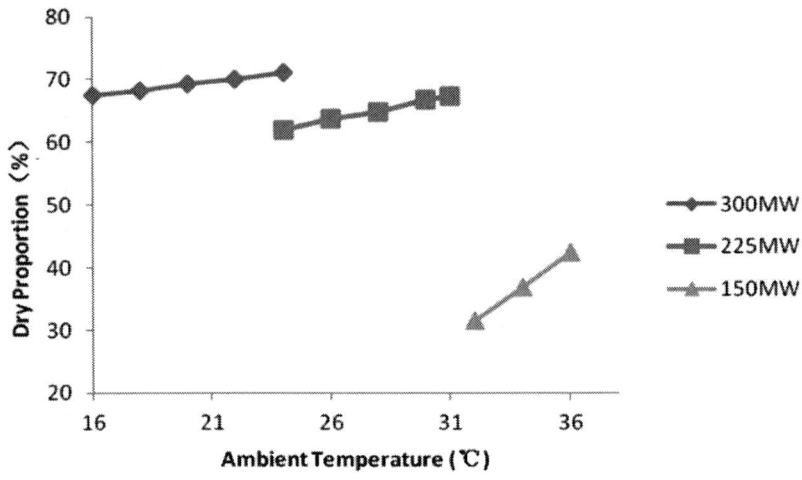

Figure 13: Power load and ambient temperature influence on dry propotion — D&Wi.

Figures 12-13 report the results of the condensing system optimization procedure for the D&WI configura-tion. The AC fan load which meet the power load at dif-fient ambient temperature is given in Figure 12. The AC fan load increses with rising ambint temperature. When the power load is 300 MW, it must open $D\&W_2$ at 24°C, and $D\&W_2$ open at 31°C when power load is 225 MW. When power load is lower than 150MW, it's no need to open $D\&w_2$.

Figures 14-15 report the results of the condensing system optimization procedure for the $D\&W_2$ configura-tion. The AC fan load which meets the power load at different ambient temperature is given in Figure 14. The AC fan load increses with rising ambint temperature. When the power load is 225 MW, the exhaust pressure can keep 15 KPa at any ambient temperature. But when the power load is 300 MW, the exhaust pressure can not keep 15 KPa at ambient temperature is above 28°C.

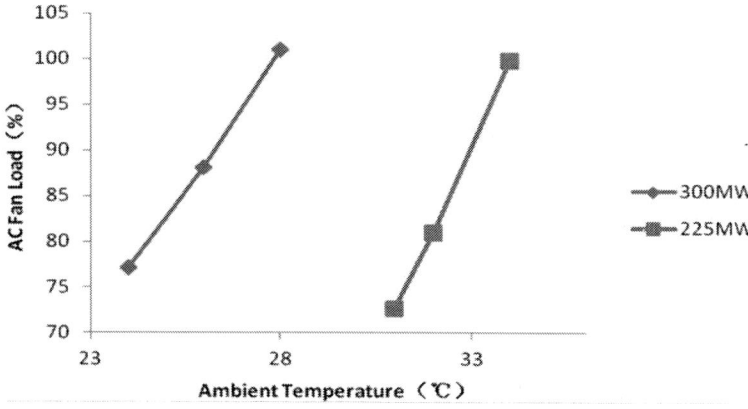

Figure 14: Power load and ambient temperature influence on AC' fan load — D&W.

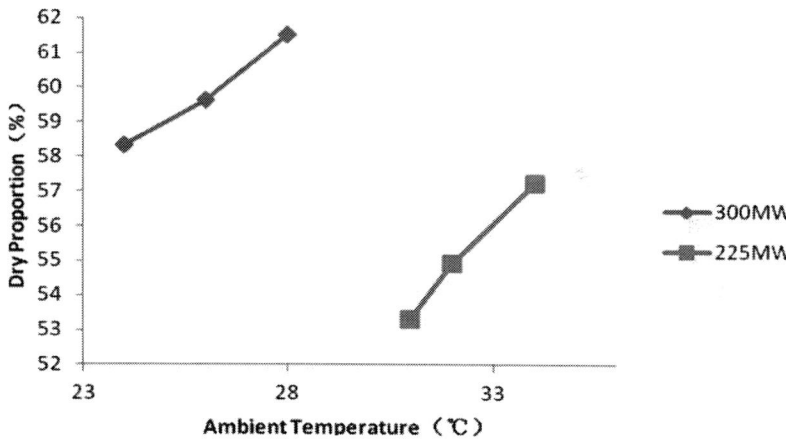

Figure 15: Power load and ambient temperature influence on dry propotion — D&W2.

CONCLUSIONS

A detailed simulation of a wet and dry cooling system installed in a steam power plant was developed and some conclusions were made as follows:)

- A parametric analysis was carried out in order to check the influence of ambient temperature, relative hu-midity, exhaust steam flow rate and air condenser fan load on the thermodynamic performances of a power plant with dry and cooling system.
- In dry and wet cooling system, the exhaust pressure decreases with rising AC fan load, and increases with rising relative humidity and exhaust steam flow rate. The effect becomes more and more appreciable with increas-ing ambient temperature.
- The heat load distribution of dry and wet cooling system in different operation situations was well devel-oped. Steam flow rate to AC decreases with increasing ambient temperature. and increases with increasing air cooling condenser fan load and relative humidity. but when the exhaust steam flow rate increases, steam flow rate to AC may increase or may be decrease decided by the ambient temperature conditions.
- The air cooled condenser resulted the best way to reject heat if the temperature is lower than 115*C, 24t . 31X.'. when the power load is 300 MW, 225 MW, 150 MW. At higher ambient temperature, the condensation should exploit the cooling capacity of the tower as much as possible while discarding the remaining heat in the as condenser.

REFERENCES

1. E. Ding, "Air Cooling Technology in Power Plant," M, Beijing: China Water Power Press, 1992 (in Chinese).
2. J. A. Heyns, "Performance Characteristics of an Air-cooled Steam Condenser Incorporating a Hybrid (dry/wet) Dephlegmator," 4 Depart of Mechanical En-gineering, Uniwersity of Stellenbosch, 2008.

3. P. Lindahl and R.W. Jameson, "Plume abatement and water conservation with wet/dry cooling towers," CTI Journal, Vol. 14, No. 2, 1993.
4. D. G. KrOger, "Air-cooled Heat Exchange and Cooling Towers," USA: Penwell Corp, Tulsa, 2004.
5. F. W. Yu and K. T. Chan, "Application of Direct Evaporative Coolers for Improving the Energy Efficiency of Air-cooling Chillers,' Journal of solar Energy Engineer-ing, Vol. 127, No. 3, 2005, pp. 430433. doi90 1115/i 1866144
6. L. X. Mott, et al, "Study on Variable Condition Features for 300 MW Direct Air-cooling Unit," J, Proceedings of the CSEE, VoL 27, No. 17, 2007, pp. 78-82.
7. S. P. Fisenko and A.I. Petruchik, "Toward to the Control System of Mechanical Draft Cooling Tower of Film Type,' Int J Heat Mass Transfer, 2005, Vol 47, pp. 31-35. doi:10.10161.ijheatmasstransfer2004.08.002
8. J. Piskorowski, R B. G. Beekett', "Condenser Perofr-manee Test and Baek-Pressure 1mProvement," EPRI, CS.5729, Vol. 4, 1988, pp. 19-23.
9. A A. Brin, A. I Petruchik, S. P. Fisenko , "Mathematical Modeling of Evaporative Cooling of Water in a Me-chanical-draft Tower," Journal of Engineering Physics and Thermophysics, Vol. 75, No. 6, 2002, pp. 1332-1338. doi.10 1013,A.1011110809044.

Citations

CHAPTER 1

Mark Ho and Melinda Hodkiewicz, "Factors That Influence Failure Behaviour and Remaining Useful Life of Mining Equipment Components," Advances in Mechanical Engineering, vol. 2013, Article ID 913048, 9 pages, 2013. doi:10.1155/2013/913048

CHAPTER 2

P. Giri, "Effort Estimation for Design Activity in Power Plant Equipments," Journal of Software Engineering and Applications, Vol. 5 No. 12, 2012, pp. 1001-1007. doi: 10.4236/jsea.2012.512115.

CHAPTER 3

Vasanthakumar B, Ravishankar H, Subramanian S (2012) A Novel Property of DNA – As a Bioflotation Reagent in Mineral Processing. PLoS ONE 7(7): e39316. doi:10.1371/journal.pone.0039316.

CHAPTER 4

Podzharov, E.. , Gálvez, J. and Sanchez, J. (2014) Acoustical Design of an Electrical Emergency Plant Using Sea Method. Journal of Environmental Protection, 5, 327-332. doi: 10.4236/jep.20J14.54035.

CHAPTER 5

Liu, W. , Xu, G. and Yang, Y. (2014) Thermo-Dynamical Analysis on Electricity-Generation Subsystem of CAES Power Plant. *Journal of Power and Energy Engineering*, **2**, 729-734. doi:10.4236/jpee.2014.24097.

CHAPTER 6

Yong Hu, Ji-zhen Liu, De-liang Zeng, Wei Wang, and Ya-zhe Li, The Optimal Steam Pressure of Thermal Power Plant in a Given Load, doi: 10.4236/epe.2013.54B054.

CHAPTER 7

A. Rastogi and H. Gabbar, "Practical Implementation of Safety Verification in LNG Production Facilities,"Open Journal of Safety Science and Technology, Vol. 1 No. 2, 2011, pp. 4359. doi:.10.4236/ojsst.2011.12005.

CHAPTER 8

C. Raghu Kumar, Sunil Tripathy, and D.S. Rao, Characterisation and Pre-concentration of Chromite Values from Plant Tailings Using Floatex Density Separator, Journal of Minerals & Materials Characterization & Engineering, Vol. 8, No.5, pp 367-378.

CHAPTER 9

Sawan, Z. (2014) Cottonseed yield and its quality as affected by mineral fertilizers and plant growth retardants. Agricultural Sciences, 5, 186-209. doi: 10.4236/as.2014.53023.

CHAPTER 10

J. Berg and S. Morling, "Process Adaption and Modifications of a Nutrient Removing Wastewater Treatment Plant in Sri Lanka Operated at Low Loading Conditions," *Materials Sciences and Applications*, Vol. 4 No. 5, 2013, pp. 299-306. doi:10.4236/msa.2013.45038.

CHAPTER 11

A. Abouzeid and A. Khalid, "Mineral Industry in Egypt-Part I: Metallic Mineral Commodities,"*Natural Resources*, Vol. 2 No. 1, 2011, pp. 35-53. doi:10.4236/nr.2011.21006.

CHAPTER 12

Tao Tang, Jian-qun Xu, Sheng-xiang Jin, and Hong-qi Wei, Study on Operating Characteristics of Power Plant with Dry and Wet Cooling Systems, doi:10.4236/epe.2013.54B126.

Index

A
Adenosine triphosphate (ATP) 171

B
Banded Iron Formation (BIF) 231
Boiler Feed-Water Pump Turbine (BFPT) 96

C
Chrome Ore Beneficiation (COB) 136, 137
Compressed Air Energy Storage (CAES) 79, 90

D
Design Activity 27, 30, 43

E
Emergency Shutdown Systems (ESS) 107
Environmental Protection Agency (EPA) 110
Ethylenediaminetetraacetic acid (EDTA) 157

F
Floatex density separator (FDS) 138
Fresh air 68
Front-end loaders (FELs) 5

G
Goldsulphides formation 240

H

Health and Safety Executive (HSE) 109
Heat Exchange Institute (HEI) 32
Heat Exchanger Unit (HEU) 27
Heat Exchanger Unit (HXE) 30

I

Independent and identically distributed (IID) 8
Indole-3-acetic acid (IAA) 154
Inherent reliability 2

L

Least significant difference (LSD) 161
Lines of Codes (LOC) 41

M

Maintenance and repair contracts (MARCs) 21

N

Nicotinamide adenine dinucleotide phosphate (NADP) 171

P

Plant growth retardants (PGRs) 155
Probability of Failure on Demand (PFD) 128
Program 32, 41
Proportional hazards assumption (PHA) 12
Proportional hazards modelling (PHM) 10
Proportional hazards model (PHM) 3

R

Rock impact hardness number (RIHN) 6

S

Safety Instrumented Systems (SIS) 107
Safety integrity level (SIL) 112
Safety related systems (SRS) 112
Software lifecycle 29
Software Quality Assurance (SQA) 29
Statistical energy analysis (SEA) 67
System analysis 34

T

Tectono-magmatic cycle 228
Total Risk Associated (TRA) 129